THE CONDITION OF SUSTAINABILITY

The idea of sustainable development is now widely used to legitimate, uphold and promote a variety of different agendas and activities, yet its practical use is increasingly being questioned and the concept itself regarded as clichéd.

The Condition of Sustainability advances both the theory and practice of this important concept, by incorporating key elements of contemporary social theory into sustainability debates, both at a conceptual and practical level. The authors begin by considering why most current approaches to sustainable development have proved inadequate, arguing that too many approaches concentrate on the definition and policing of various sustainability limits, and a more dynamic conception is necessary. Case study material, used for illustration and analysis throughout the book, focuses on the food system, particularly the sugar industry in Australia and Barbados – sugar being the most widely produced agricultural commodity in the world and widely associated with a range of environmental and social impacts which can be regarded as unsustainable.

Bringing together key ideas from social theory, food regimes and sustainability debates, *The Condition of Sustainability* explores the political economy of sustainable development and presents a new and powerful way of thinking about sustainable development as well as a methodology for applying these ideas. It is now all too easy to dismiss the concept of sustainability out of hand; this book demonstrates that such dismissal is, at the very least, premature.

Ian Drummond is Learning Support Officer, Department of Geography, University of Hull; **Terry Marsden** is Professor of Environmental Policy and Planning, Department of City and Regional Planning, Cardiff University.

Routledge Research Global Environmental Change Series

THE CONDITION OF SUSTAINABILITY

Ian Drummond
and
Terry Marsden

Global Environmental Change Programme

London and New York

First published 1999 by Routledge
11 New Fetter Lane, London EC4P 4EE

Simultaneously published in the USA and Canada by Routledge
29 West 35th Street, New York, NY 10001

© 1999 Ian Drummond and Terry Marsden

The right of Ian Drummond and Terry Marsden to be identified as the
Authors of this Work has been asserted by them in accordance with the
Copyright, Designs and Patents Act 1988

Typeset in Garamond by RefineCatch Ltd, Bungay, Suffolk
Printed and bound in Great Britain by
Biddles Ltd, Guildford and King's Lynn

British Library Cataloguing in Publication Data
A catalogue record for this book is available from the British Library

Library of Congress Cataloguing in Publication Data

Drummond, Ian,
   The condition of sustainability / Ian Drummond and Terry Marsden.
       p.   cm.—(Global environmental change series)
   Includes bibliographical references and index.
   ISBN (invalid) 0–415–19693–8 (hc.)
   1. Sustainable development.   2. Sugar trade—Australia—Case
studies.   3. Sugar trade—Barbados—Case studies.   I. Marsden,
Terry.   II. Title.   III. Series.
HC79.E5D78   1999
338.9—dc21   98–24232
                                                                         CIP

ISBN 0–415–19493–8

CONTENTS

FIGURES

TABLES

ABBREVIATIONS

ABARE	Australian Bureau of Agricultural Resource Economics
ACP	African, Caribbean and Pacific
BADC	Barbados Agricultural Development Corporation
BAMC	Barbados Agricultural Management Corporation
BCSA	British Commonwealth Sugar Agreement
BET	Basic Export Tonnage
BLP	Barbados Labour Party
BNB	Barbados National Bank
BSES	Bureau of Sugar Experiment Stations
BSIL	Barbados Sugar Industry Limited
BS&T	Barbados Shipping and Trading Company
BWU	Barbados Workers Union
CLICO	Caribbean Life Insurance Company
CSA	Commonwealth Sugar Agreement
CSR	Colonial Sugar Refiners Limited
DLP	Democratic Labour Party
DPI	Department of Primary Industries
ECU	European Currency Unit
EEC	European Economic Community
EU	European Union
FAO	Food and Agriculture Organisation of the United Nations
GOB	Government of Barbados
HFCS	High Fructose Corn Syrup
HIP	Heavily Indebted Plantation
IMF	International Monetary Fund
ISA	International Sugar Agreement
ISO	International Sugar Organisation
IUCN	International Union for the Conservation of Nature
MIP	Moderately Indebted Plantation
RAS	Rural Adjustment Scheme
SCI	Sparks Companies Incorporated
SCIST	Senate Committee on Industry, Science and Technology

ABBREVIATIONS

tc	tonnes of cane
ts	tonnes of sugar
ts/y	tonnes of sugar per year
UNEP	United Nations Environment Programme
WCED	World Commission on Environment and Development
WWF	World Wide Fund for Nature

INTRODUCTION

Seemingly simple, intuitively rational and self-evidently expedient, sustainable development is a notion which engenders an instant and almost instinctive attraction. As Redclift (1992:1) suggests, 'like motherhood and God it is difficult not to approve of it'. Since it first achieved popular recognition following the World Commission on Environment and Development (WCED) in 1987, sustainable development has rapidly become 'the watchword for international aid agencies, the jargon of learned planners, the theme of conferences and learned papers, and the slogan of environmental and developmental activists' (Lélé, 1991:607). Indeed, the concept is now widely used to legitimate, uphold and promote a wide variety of different agendas and activities. However, many analysts have come to regard it as an insubstantial and clichéd platitude unworthy of further interest or research, and, perhaps even more significantly, theorising of the idea seems to have reached something of an impasse. Whilst it now seems all too easy to dismiss the concept out of hand, we believe that this would be premature. Certainly, we cannot dismiss the material dysfunction and moral unacceptability of present day development so readily. And, importantly, what we must not do, is to reject the idea of sustainable development because we have failed to address the key conceptual and methodological challenges which it presents. This is not a worthy basis for rejection. If we do this, we tacitly accept all that is unsustainable.

This volume is an attempt to move both the theory and practice of this important concept beyond the impasse at which it is currently stalled. After a decade when social scientists have been keen to identify themselves with the concept, and to point to the need to understand the human interactions involved in environmental change, it is now increasingly clear that neither global systemic or rational technicist approaches are in themselves adequate for evolving a progressive agenda. This problem is represented in two ways. First, more effort is needed in understanding the real potential and existing constraints in achieving sustainability goals grounded as they are in the social and natural environments in which people are situated. This needs to take us beyond the definition of environmental limits and capacities or the

1

abstract attribution of cost criteria to contingent natural and social conditions. Additionally, if we are to progress beyond abstract definitional arguments, we need to appreciate the construction of the gaps which exist between real conditions and potentially feasible sustainable scenarios. Second, it is becoming increasingly recognised that the current status of sustainability debates in the social sciences represents an explicit methodological challenge in terms of empirical and analytical approach. In a vertical sense, we now see sustainability discourses becoming embedded in institutional structures, while horizontally, we witness the uneven articulation of sustainability problems. For social scientists this means that there have to be more robust methodological techniques and ways of incorporating sustainability perspectives into existing theoretical and methodological frameworks. These need to go beyond the simple realisation and recognition of global unsustainability and environmental risk. The need is to develop and integrate more effectively social theory and concepts with questions of nature and sustainability.

Within this agenda, the arguments presented in this volume represent one attempt to develop a social sustainability perspective for the purpose of gaining an improved understanding of environmental uneven development. Its contribution lies in incorporating some key elements of contemporary social theory into sustainability debates (Redclift and Benton, 1994) both at a conceptual level and at the level of comparative empirical enquiry and analysis. It is argued that to really progress sustainability objectives it is necessary to understand the structural and regulatory causes of current unsustainable practices. We develop this agenda by employing a combined regulationist and realist approach to the case of one global food sector: sugar. While realist and regulationist debates have largely been silent on the course of the sustainability debates of the past decade (see Harvey, 1996; Marsden *et al.* 1996), it is argued and demonstrated that a realist understanding of causality informed by insights from regulation theory allow an important bridge to be built between current dualisms in the sustainability debates associated with a focus either on structural and global phenomena, or its obverse, the actual mechanisms and tools needed to improve upon the unsustainable. After outlining these theoretical perspectives and the gaps and potentialities they embody, the book details how this modified theoretical approach can be applied in order to elaborate a new methodological agenda which can deal with the temporal, but particularly the spatial comparative dimensions such a perspective suggests. This goes beyond current debates about globalisation and environmental change (see Yearley, 1996) and begins to develop a framework which provides theoretical and empirical comparative analysis of uneven environmental development. In this regard, while the empirical reference point is the food system and within this the contrasting sugar industries of Australia and Barbados, the text intends to provide a broader theoretical and methodological window on (i) the scholarly trajectory of the sustainability perspective and (ii) the potential application of the derived approach to the study of sustainable food and rural development more generally.

The volume starts by considering why most current approaches to sustainable development have proved to be inadequate. We argue that this reflects both a lack of depth and inappropriate direction in thinking about sustainability. Our central argument is that most existing approaches to sustainable development have focused pragmatically, but we would argue inappropriately, on attempting to objectify what is and is not sustainable in particular places at particular times. Sustainability and unsustainability have, however implicitly, been seen as concrete and absolute. Thus attention has been concentrated on different forms of sustainability 'metrics', and subsequently on the most effective means of policing the limits defined by these metrics. In this book we argue that such approaches are limited in their potential because they focus on the question of *where the line should be drawn and how it might best be policed*. This is at best a secondary question. The more incisive question, and the one we attempt to address in this volume, is *why is it that the line will always tend to be crossed wherever it is placed?* In attempting to address this second question, we use a realist ontology to explore the political economy of sustainable development. We suggest that rather than thinking in terms of concrete definitions of what is and is not sustainable, it is more useful to construct a dynamic conception within which the key concern is the sustainability or otherwise of development trajectories. Within this perspective, it becomes possible to explore the ways in which the nature of development is *conditioned* by particular social institutions and values. Understanding specifically why and how development tends to be conditioned in ways which make unsustainable outcomes the norm is, we argue, the key to understanding how more effective approaches to sustainable development might be formulated.

Chapter 1 provides a critical review of both the theory and practice of sustainable development. The discussion builds on a critique of existing approaches to sustainability and begins to define a new agenda. It is argued that attempts to 'objectify' what is or is not sustainable are ill-conceived and that progress towards sustainability requires a more dynamic conceptualisation of sustainability with particular examples of 'unsustainability' being understood and addressed as outcomes of social processes. It is argued that one way forward lies in the development of a multi-level but unified explanation of the causality of unsustainable events and practices. Specifically, it is suggested that a critical realist approach, allied to insights from regulation theory, can provide a deeper and potentially useful understanding of how and why unsustainable events tend to be the norm.

Chapter 2 demonstrates how a realist ontology can provide a powerful conceptual basis for understanding the causes of unsustainable practices and events. Beginning with a review of modern conceptions of realism, the chapter then discusses how these define a methodology relevant to advancing an understanding of sustainability issues and how they can form the basis of new forms of regulation. From a realist perspective unsustainable practices and events are seen as outcomes rather than as discrete or purely contingent events and this has important policy implications. A realist interpretation of the causes of unsustainability suggests a range of possible points for intervention, and our argument, which is expounded through this volume, is that a particular significance needs to be attached to the conditions which selectively activate structurally defined causal mechanisms. The final section of this chapter outlines the methodology used to construct a realist explanation of unsustainable patterns of development in and around the Barbadian and Australian cane sugar industries. Consideration is given first to the general methodological considerations posed by a realist approach and subsequently to the specific research methods used in the case studies presented in chapters 6 and 7. Although it is recognised that established realist methodology is often somewhat idealised and difficult to apply in practice, it is argued that these difficulties are not insurmountable.

Chapter 3 begins by outlining developments in regulation theory. Consideration is then given to the ways in which modes of social regulation influence patterns of development. In particular, it is suggested that current modes of social regulation selectively legitimate and empower strategies which sustain existing social formations. They do this by translating the contradictions which emerge within these into materially and morally significant forms of unsustainability. Unsustainability thus becomes the outcome of particular modes of social regulation and their attempts to maintain particular social formations. The final section of the chapter considers how such an understanding of the role of regulation in capitalist societies can inform thinking on sustainable development.

Chapter 4 applies the ideas developed in the previous chapter to the international food system. The chapter begins by outlining the historical development of 'food regimes' in the modern capitalist system since 1860, summarising different phases of regulation: first, second and third food regimes, the changing production–consumption relationships involved in these phases and their significance for uneven development and agricultural production. Consideration is then given to the relationship between regulation, the emergence of particular social formations and sustainability.

Chapter 5 builds on the analysis of food regimes in chapter 4 and applies the general ideas developed there to the sugar sector. Sugar is the most widely produced agricultural crop in the world. It is also a crop which is closely associated with a range of environmental, economic and social practices and impacts – many of which are considered to be 'unsustainable'. The chapter first briefly describes the characteristics and history of sugar production. Consideration is then given to current patterns of production and consumption. Key features of the global sugar economy, including the ACP

Sugar Protocol of the Lomé Convention, are outlined in the final section of the chapter. This provides a context for the case studies of sugar production in Barbados and Australia which form the next two chapters of the volume.

Chapter 6 begins with a brief description of Barbados and the island's history. The extent of the crisis which has befallen the sugar industry is then outlined. The remainder of the chapter is concerned with understanding why the industry has collapsed and how this unsustainability is related to a wide range of other unsustainable practices and events on the island. We then consider a range of explanations which might account for the collapse of the sugar industry. What emerges here is a complex picture composed of partial and often contradictory explanations confused by the biases and self-interested perceptions of many of those involved in the industry. Within this, however, it is clear that unsustainability in present day Barbados needs to be understood in relation to the island's racial and class structures, and in relation to the resulting contradictions and tensions which exist within Barbadian society. More generally, consideration is then given to how this unsustainability is related to a range of other unsustainable outcomes. The discussion focuses on the strategies which have been adopted by the island's elite group to sustain their own status and privilege and the institutional and social context which has legitimated and empowered these strategies. The final section of the chapter attempts to interpret development in Barbados in explicitly realist terms using the methodological approach outlined in the first section of this volume.

Chapter 7 begins with a brief description of Australian sugar producing areas and the evolution of the Australian sugar industry. The current structure of Australian sugar production is then outlined in some detail. Particular attention is paid to the highly structured nature of the sugar industry regulatory system which has been in place for most of the twentieth century. It then considers the impacts of the deregulatory process currently being enacted in two sugar producing regions in Queensland. The discussion focuses on the problems currently faced by the industry and the coping strategies which have been adopted. Particular attention is paid to the range of environmental, agronomic, economic, social and moral forms of unsustainability currently occurring within the Australian industry. The second half of the chapter provides a deeper analysis of recent events in the Australian sugar industry in order to better explain both the potential for the unsustainability of the industry itself and the range of unsustainable outcomes which have been associated with the development of this sector. The discussion focuses on the contradictions and tendencies to dysfunction which have emerged within this sector and the strategies through which these have been addressed. Particular attention is paid to the ways in which the regulatory system has affected the development of the industry and the relationship between the inherent unsustainability of the traditional socio-economic formation within the sugar sector and various materially and morally unsustainable outcomes which have occurred.

Chapter 8 reconsiders the conceptual framework and approach to sustainability developed in the initial chapters of this book in the light of the results obtained from the case studies. The model linking unsustainable outcomes with real causal factors developed in Chapter 3 is re-evaluated firstly in relation to the sugar sector and subsequently in terms of national and global scale food systems. Particular consideration is given to the ways in which causal mechanisms are selectively 'activated' by modes of social regulation and the ways in which particular social formations 'condition' the nature of development. The final sections of the volume evaluate the wider practical utility and policy relevance of the approach to sustainability developed throughout the book and consider how this perspective will allow sustainable development to be articulated in the general case. The discussion here will address both the conceptual and empirical validity of the approach and the methodological challenges facing scholars who adopt either political economy or more social constructivist perspectives on sustainability. The chapter concludes by suggesting how the approach to sustainability developed throughout the book might be further tested, refined and progressed.

1

SUSTAINABLE DEVELOPMENT
The impasse and beyond

A good idea? Contested interpretations and utilities

The most widely known and used definition of sustainable development is that provided by the World Commission on Environment and Development (1987:8) which suggests that sustainable development is 'development which meets the needs of the present without compromising the ability of future generations to meet their own needs'. This definition says very little about what sustainable development actually is or how it might be achieved. Since 1987, there has been a plethora of attempts to define the concept more closely, and explicitly or implicitly, it has been interpreted and reinterpreted in different and often markedly discordant ways.

To some extent the increasingly widespread use of notions of sustainable development may well reflect the inherent rationalism of the idea. Certainly, images of the unsustainable conjure up affective visions of Malthusian crisis and catastrophe. However, as Adams (1993:218) suggests, 'sustainable development is a flag of convenience under which many ships sail, and it is this catholic scope that goes a long way to explain its power and popularity'. In practice, the inherent ambiguity of the idea is often exacerbated by the fact that a range of terms such as 'sustainable development', 'sustainability', 'environmental sustainability', 'sustainable growth', etc., are used more or less interchangeably when in fact they are held to have specific and significantly different connotations. As English Nature point out:

> It is important to distinguish between 'sustainable development' and 'sustainability'. Sustainable development is a broader social objective: it is concerned not just with environmental protection but with the achievement of other social objectives. . . . This is not the case, however, for 'sustainability'. This is concerned only with the environment, and it can be defined quite precisely. (It is true that sometimes the adjective 'sustainable' refers to social and political sustainability as well as to environmental – some authors have argued that the stability and durability of social institutions as much

7

as the environment are necessary to a 'sustainable society'. But 'sustainability' has come to be almost exclusively an environmental term.) Thus 'sustainability', in a narrow sense, is related to the resilience of ecosystems, that is their ability to withstand various types of stress, rather than any social or economic considerations.

(English Nature, 1992:17)

Whilst recognising that a broad spectrum exists between these two polarisations, Hodge and Dunn (1992:16) have summarised the essential characteristics of 'hard' and 'soft' interpretations (see table 1.1). Whilst English Nature (1992:16) argue that 'it is not impossible to reconcile these two positions: sustainability constraints can be applied to some aspects of the environment while others are traded off in terms of costs and benefits', it is far from clear whether this is indeed the case. Although there is no single uncontested definition of what sustainable development is, nor any consensus about how it might best be achieved, a range of themes and issues, including needs, equity, intergenerational equity and resources, are common to most interpretations.

One key aspect of sustainable development which stems directly from the Brundtland Commission definition is that of intergenerational equity. This is normally taken to mean that the global resource base should not be degraded in ways which deprive future generations of the ability to attain a level of well-being equivalent or superior to that achieved today. This has relatively straightforward implications for the ways in which flow or continuous resources are exploited, but becomes more problematic when policy regarding stock resources is considered (see, for example, Rees, 1990). Moreover, in both of these cases, the position becomes much less clear when possible advances in technology are included in the analysis. Environmental

Table 1.1 Alternative positions on sustainable development

Soft sustainability	Hard sustainability
Prevention of catastrophe for human society	Promotion of society in harmony with ecosystem
Acceptance of science and modern technology	Questions science; seeks alternative technology
Anthropocentric	Ecocentric
Intergenerational distribution treated separately	Intergenerational distribution integral to sustainability
Lower environmental risk aversion	High environmental risk aversion
Marginal changes to existing systems and institutions required	Shift to new systems and institutions

Source: Hodge and Dunn, 1992.

economists tend to argue that it may be desirable to exploit natural resources in ways which degrade the overall natural resource stock, providing losses of 'natural capital' can be substituted for by future developments in 'human capital' (Barbier, 1989; Pearce and Turner, 1990). However, this position assumes the potential for incremental gains in the utility of available technology, and beyond this, it pays scant regard to the not indefensible contention that far from being a panacea, technology is a key factor underlying many of the world's contemporary problems.

To some extent, concern for future generations appears to have diverted attention away from consideration of the lack of equity manifest in present day patterns of development. To many analysts, 'development' is a moral concept which implies both the maximisation of well-being and the progressive achievement of equality in society. From this perspective, present day patterns of uneven development are both morally unsustainable and a barrier to the achievement of more sustainable patterns of development in the future (see Smith, 1984). However, bar a few cases (see, for example, Redclift 1987; O'Connor, 1988), the significance of uneven development to sustainability issues has yet to be adequately considered.

Almost all definitions of sustainable development place a central significance on the role of resources and many visions of unsustainability are founded on the contention that resources are being degraded or destroyed. Although she is clearly critical of these neo-Malthusian conceptions, Rees also points out that an overly cornucopian outlook is equally insupportable:

> Resources cannot be defined in physical terms, nor can scarcity be regarded as a problem in any narrowly physical sense. It is now largely accepted that in the foreseeable future economic development will not be brought to a catastrophic halt as it hits the stock resource availability barrier. Nor does it appear likely that market imperfections, geopolitical problems or environmental controls will create any really significant mineral scarcity problems for the now advanced nations. However, it is not possible to be so sanguine either about the future for countries in the Third World or about the continued availability in all societies of environmental resources.
>
> (Rees, 1990:58)

As Rees suggests, access to resources is unequal and spatially differentiated and it seems likely that a range of 'environmental resources' are being modified in ways which prejudice future development not just in the South but throughout the world. From this perspective, it is perhaps more useful to consider those technical, economic and social processes which serve to define and redefine particular resources and the resource base as a whole. Thus an understanding of the contexts and processes which underlie the overexploitation of resources may be crucially significant. As Harvey explains:

'Resources' can only be defined in relationship to the mode of production which seeks to make use of them and which simultaneously 'produces' them through both the physical and mental activity of the users. There is, therefore, no such thing as a resource in the abstract or a resource which exists as a 'thing in itself'.

(Harvey, 1977:226).

This clearly is the case but, as Moore (1993:396) points out, 'to say that resources are socially produced and culturally constituted does not, however, make them any less real or material'. Neither does this lessen the material significance of the unsustainable practices and events which tend to be associated with the dynamic definition and redefinition of resources. What this does suggest, however, is that however concrete they may appear, unsustainable practices and events cannot be fully understood outside the social and economic contexts in which they occur.

Sustainability debates thus need to move beyond the somewhat naïve conceptions of a 'resources crisis' engendered by publications such as 'Limits to Growth' (Meadows *et al.*, 1972). However, whilst the need for a more sophisticated understanding of resources, which fully accepts that these are fundamentally socially defined, is widely recognised (see for example, Rees, 1990), many current conceptions of sustainable development still adopt an approach not so very far removed from that of the neo-Malthusians during the 1970s. Currently prevalent concepts such as 'maximum sustainable yield', 'carrying capacity', 'critical loads' or indeed the idea of 'trade-offs' all imply the existence and significance of materially defined limits. Owens (1994), for example, sees a central role for limits defined by 'critical natural capitals' within the planning process. Healy and Shaw (1994) use the term 'capacity' in a similar context. The requirement for such limits may appear to be intuitively obvious and a practical necessity, but the utility of any approach to sustainable development which centres solely on such metrics is highly questionable.

Redclift (1991:7) suggests sustainable development may be about 'meeting human needs, or maintaining economic growth or conserving natural capital, or about all three'. But the whole point of sustainable development, the only point which differentiates the concept from narrower ideas such as environmentalism, is that the concept is more than the sum of its parts. It is not just a multi-dimensional concept, it is *fundamentally integrative*. The key problem, however, is that it is profoundly difficult to grasp this multi-dimensionality. Most current approaches focus on and privilege a particular dimension, be it economic, environmental or social, and what results is often something less than sustainable development. Sustainable development will remain little more than rhetoric unless it can be used to inform policy in objective ways, its credibility is impeached by any approach which through partiality or prioritisation implicitly reduces the concept to something less than that which a properly holistic conception requires. However, most existing approaches to sustainable development have either abandoned any real claim to be integrative or are flawed in some other respect. The majority of these approaches fall into one of three categories: those which assume that such a state can be managed; those based on adherence to certain principles or the management of certain 'currencies'; and the model of sustainability derived from environmental economics.

Notwithstanding considerable rhetoric about value shifts and the like, the actuality of the unsustainable is material and often pressing, and it is this concrete reality which is most obvious and most readily addressed. When one also considers the fact that sustainability policy is normally constructed and enacted within discrete politically defined spaces, and under pressure to produce results, it is perhaps not surprising that, in practice, most attempts to promote sustainable development have involved concrete measures designed to prevent or control specific aspects of development directly. In practice, this has normally involved strategies designed to define and subsequently police some form of 'sustainability limits'. Any such approach is likely to encounter at least four problems.

First, it is profoundly difficult to determine where such limits should be set in a truly objective manner. Not least because optimal limits will change through both space and time. A good example of this type of problem is provided by attempts to define 'total allowable catches' within the European Union's Common Fisheries Policy (CFP). Population dynamics within fish stocks are complex, subtle and subject to natural fluctuations. This makes any objective determination of appropriate sustainability limits difficult. It also makes any definition easily contestable and difficult to police (Drummond and Symes, 1996).

Second, any determination of physical limits is not very meaningful outside the social context in which it occurs:

> limits are not, however, set by the environment itself, but by technology and social organisation. Physical sustainability cannot be secured without policies which actively consider access to resources and the distribution of costs and benefits.
>
> (Adams, 1993:211).

This precisely has been a key problem for the CFP, the difficulties encountered in attempting to regulate fishing effort within EU waters have proved to be effectively insurmountable. At least, attempts at regulation based mainly on total allowable catches have proved to be an almost total failure. The various national fisheries involved have different traditions and occur at different scales ranging from local, family-based enterprises to highly capitalised, large scale commercial fishing companies. Thus the difficulties involved have included not just the need to formulate policy under conditions where the efficiency of technology has increased progressively and where these technologies have been differentially adopted, but also where there is differential ability to access resources and differences in perception of what constitutes the fishery resource. This single resource is viewed very differently by small fishing communities and the large commercial fishing companies.

A third problem relates to the manner in which policy is articulated. One question here relates to the most appropriate scale for intervention. In practice, there are a wide range of prescriptions regarding the most appropriate scale at which the concept might be operationalised. One perspective on this problem is adopted by Gardner (1990:337), who argues for a bottom-up approach in which decision making at the community level provides the framework for 'achieving development to meet the needs/aspirations of the local population, respects cultural diversity, and maintains ecological systems'. Not all approaches focus on this scale however. Nijkamp and Soetemann (1988:626) use the term 'area management'. Norgaard (1988) argues for 'regionally sustainable systems'. The *Caring for the Earth* report (IUCN *et al.*, 1991) advocates a nationally based approach with each region being treated as an integrated system within which the carrying capacities of these systems and the needs of indigenous populations can be used as the basis for policy formation. A nationally based framework would, to some extent at least, address the problem that otherwise suitable regions may well not be congruent with relevant institutional structures. Notwithstanding the role of supra-national governments and agencies, the national and supra-national state remains the locus of most regulation.

There are, however, a number of general difficulties in articulating sustainable development within any discrete spatial framework. As Dovers and Handmer (1992:262) suggest, 'the task nations face will be compounded by the fact that their own particular challenges, in more and more cases, can no longer be dealt with in isolation but must be placed in the context of global environmental change and the global economy'. Certainly, it seems to be unlikely that sustainable development can be achieved with discrete, small scale units. At least not if that development involves meeting the material needs of the population. Most of the features of development which we consider to be vital in that they improve human well-being – health care, education, food security, etc. – are dependent on complex, if not globalised, systems of production and distribution. This complexity does not simply reflect the fact that individual regions are invariably dependent on external linkages, even more problematic is the fact that the sustainability of society as a whole is ultimately dependent on the sustainability of its constituent elements. Development – in the sense of increasing human well-being – is not threatened by the collapse of any single enterprise, but it must be dependent on the cohesive aggregation which the individual elements constitute. In practice, the sustainability of society as a whole is dependent on some form of 'structured coherence' which cannot be fully anticipated or managed (see, for example, Harvey, 1989).

Narrowly focused approaches to sustainable development require the ability to regulate effectively in a detailed, co-ordinated and integrated manner throughout the whole gamut of human activity. However, as Hayek (1988) suggests, contemporary societies and economies are so complex that adequate knowledge for large scale planning cannot possibly exist. Indeed, the idea of planning for sustainability, at least in the sense that it is normally understood and translated into practice is, in itself, insufficient to produce the desired results (Meadowcroft, 1997). Certainly, it seems to be highly improbable that sustainable development can be effectively promoted solely through use of concrete forms of regulation which address specific problems, be they prohibitive legislation, fiscal measures, or whatever. Problem definition in these cases tends to inadequately incorporate social action and process as dynamic and potentially reproducible forces. Such an agenda is flawed in its conception, and is probably untenable. The effective articulation and operationalisation of such an approach is almost certainly beyond the scope of human agency. It would require the management of what is in practice probably unmanageable.

A fourth, and perhaps even more significant problem, is the fact that conceptions of sustainable development which begin from the premise that limits need to be determined and then policed, predicate particular and arguably ill-conceived approaches to the promotion of sustainable development. Although the totality of development is constituted in specificity, a substantive approach to sustainable development necessarily transcends the limited scope of specific policies or measures. Thus the problem is not so much the geographical scale at which policy is articulated, but rather the nature of the policy measures being used, and underpinning these the theoretical basis of policy formation. Narrowly focused approaches designed to address what are perceived to be unsustainable events directly ask the wrong questions, provide largely irrelevant answers, and lead directly to inappropriate strategies. Such approaches are problematic because as Adams suggests, they address a target which is constantly moved by technological innovation and changes in social organisation. But this is a mere technical difficulty, it is not the real problem. In themselves, such strategies will always be inadequate because however pragmatic and apparently positive they may appear, they are palliative, they address the outcome rather than the cause of the problem. They represent an insufficient and inadequate basis for the achievement of long-term sustainability. Approaches of this type focus on the question of *where the line should be drawn and how it might best be policed*. This is, at best a secondary question. The question which we should be asking is *why the line will always tend to be crossed wherever it is placed.*

There are then clear and significant problems associated with any approach to sustainable development predicated around the idea of sustainability limits. However, a logical corollary of the suggestion that sustainability is defined by limits is that any truly sustainable mode of development must maintain some form of equilibrium. It may be possible to maintain the viability of other types of systems in the short term, but in the end, non-equilibrated systems cannot be sustainable. For example, fishing effort which exceeds an ecologically determined maximum sustainable yield can be sustained by increasing effort or through the use of subsidies, but in such cases equilibrium is achieved artificially (Drummond and Symes, 1996). For development to be truly sustainable its dynamism must be internalised. No sub-system which borrows from other geographical or temporal subsystems can be sustained indefinitely. It may be possible to rob Peter to pay Paul in the short term, but this can only be a temporary strategy. That is not to say that sustainable development needs to be conceived of as some fossilised steady state system. Given that the various constituent dimensions of sustainability are by their nature dynamic, it surely follows that sustainable development cannot exist as some simple equilibrium state which can be regulated by reference to constant limits and some simple notion of balance between the various dimensions. This is not in itself a problem, systems can, and do, exist in states of dynamic equilibrium where the conformation of the system changes but within which an essential balance is maintained.

In one sense, a truly sustainable system would be one in which all processes were internalised by virtue of what the system was. As Pierce suggests, 'in the end sustainable development will be a self-enforcing process capable of achieving its own equilibrium' (Pierce, 1992:318). From this perspective, sustainability would be achieved through some form of homeostasis – it would occur because of the nature of the mode of development rather than through any form of environmental management or social policy. In such a system sustainable development would be normative. However, whilst it is sometimes described as such (see for example Pearce, 1995:9) sustainable development, in itself at least, is clearly not a normative theory. Most certainly, it is not a normative theory in the sense of, for example, central place or industrial location theories, which suggest an objective and logically determinable position towards which real world patterns will, or at least ought to, gravitate. Indeed, one might well argue that while 'development' – capitalist development at least – may promote patterns which reflect certain laws or tendencies, the outcomes which tend to occur are for the most part antipathetic to notions of sustainability.

Although some form of equilibrium appears to be essential to sustainability, it is far from clear whether any conceivable system will be truly homeostatic in the sense that it maintains its own equilibrium. The problem here is that it is equally uncertain whether it would be possible to devise forms of regulation which would impose equilibrium on a non-homeostatic system. Yanarella and Levine, however, begin to outline how some form of 'homeostatic balance' might be achieved in practice:

14

Activities or processes are neither good nor bad when taken by them-selves. Instead a desired activity can take place in a larger system only by finding its balances within that more encompassing system. In order to seek such a balance, the process must have a context or system within which the balance may occur. It is thus a question of relationships. For, even when a new component introduced into the existing system of relationships upsets the balance of the larger sys-tem, a counter-tendency may be set in motion whereby the larger system responds to the change by striving for a new state of equilibrium.

(Yanarella and Levine, 1992:770)

Implicitly, at least, the problems inherent in managing sustainable de-velopment have been recognised for some time, and have given rise to a broad approach to sustainability based on the premise that whilst managing de-velopment in all its complexity is a seemingly impossible task, it may well be possible to base policy around a series of 'principles' or 'currencies'. Agenda 21, for example, promoted such an approach as, to a large extent, does UK government policy (Secretary of State for the Environment, 1994). A similar position is proposed by Dovers and Handmer who advocate the utility of a systems based approach to understanding and operationalising sustainable development, 'the first principle is the need for a systems approach: accept-ing and designing approaches which suit the axiomatic proposition that the sustainability is a whole-system problem. Sectoral or single issue approaches are clearly inadequate' (Dovers and Handmer, 1992:274). They argue that a systems approach is the 'logical place to begin' because it allows progress which could not be made through 'increasing efforts in specialisation and reductionism'. They quote Laszlo to substantiate their conviction:

A systems approach can look at a cell or an atom as a system, or it can look at the organ, the organism, the family, the community, the nation, the economy, the ecology as systems and it can view even the biosphere as such. A system in one perspective is a subsystem in another. But the systems view always treats systems as integrated wholes of their subsidiary components and never as the mechanistic aggregate of parts in isolable causal relations.

(Laszlo, 1972:14)

At face value such assertions would appear to suggest that systems theory may well have some relevance to the articulation of sustainable development. Such an approach might allow analysis which simultaneously addressed the concept at different scales. As Dovers and Handmer suggest:

Given a whole-system approach, and the absence of any hope for complete information (or for its use in much decision making even if it existed), system-wide factors and indicators should be a priority, both in terms of understanding system behaviour and identifying policy options with effective generic potential. An example here might be that, following a recognition of energy and money as basic system 'currencies', and the former as a prime indicator of societal load on the environment, energy taxation seems a logical area for exploration when assessing policy instrument choice for sustainable development.

(Dovers and Handmer, 1992:275)

The basis of this sort of approach is that whilst it may well be impossible to understand the development in all its complexity, it may nevertheless be possible to manage it albeit in a 'grey box' manner. In practice, the utility of this sort of approach is questionable. How sensible would it be to assume that something akin to 'monetarism' might promote sustainable development when history would suggest that such an approach cannot even ensure a sustainable economy? In practice, it is far from clear what currencies are appropriate or how these should be specified and managed to produce sustainable development. Similarly, the suggestion that sustainable development can be achieved through adherence to certain 'principles' appears to reduce the requirement for 'managerialism', but this is largely illusory. At least it is if the principles involved are defined in a subjective, unembedded and non-specific manner. Attempts to apply such general principles without any really adequate understanding of their detailed significance is hardly likely to be effective. The real need is to move beyond managerialism.

The one approach to sustainability which claims to achieve moving beyond the need to directly plan and manage development is that emerging from environmental economics. According to Dickens (1992:13), 'neo-liberals see the promotion of successful market economies as the principle means through which ecological and environmental problems can be solved'. As Rees puts it, proponents of the market claim 'impersonality, neutrality and freedom from political pressures' which are sharply contrasted with the:

imperfect, value-laden self-interested human beings who feature in more bureaucratic regulatory regimes . . . theoretically at least, economic incentives give producers the freedom to find the most cost-effective methods of pollution control and resource conservation, while also enabling consumers to establish the desired mix and allocation of resource goods and services. If properly programmed, markets could ensure that critical resource and environmental limits were respected, but also allow individual cultural groups to select their own sustainable development packages with variable contents.

(Rees, 1992:386)

Neo-liberal interpretations of sustainable development typically suggest that the concept involves 'maximising the net benefits of economic development, subject to maintaining the services and quality of resources over time' (Pearce and Turner, 1990:42). According to Barbier (1989), this entails identifying the optimal level of interaction between the biological, the economic and the social systems through a dynamic and adaptive process of trade-offs. Pearce and Turner similarly emphasise the importance of trade-offs between present systems and between generations; a viewpoint which leads these authors to suggest that 'the issue, then, is how we should treat natural environments in order that they can play their part in sustaining the economy as a source of improved standard of living' (Pearce and Turner, 1990:43). From this perspective, there is no imperative requirement to preserve any particular elements of the natural environment, the issue is one of when it is appropriate to substitute 'man-made capital – machines, factories, roads, – for natural capital. Indeed traditional economic growth has proceeded on this basis: machines have been substituted for animal power, electricity for fuel wood, artificial fertilisers for organic manures, and so on' (Pearce and Turner, 1990:48). Whelan (1989:29) extends this argument to what appears to be its logical conclusion by suggesting that resources should be exploited rather than conserved because 'market forces and human ingenuity will always take care of shortage by providing solutions which leave us better off than we were before'.

The problems of this approach are, however, well known. See for example, Redclift (1988), Pierce (1992:308) and Jacobs (1994) who suggests that 'in neo-classical economics there is an "atomistic–mechanical worldview" in the identification of and solution to economic problems. In theory, natural capital is divisible and indistinguishable from human-made capital', something which he suggests is one of several factors which have allowed 'economic systems to diverge from natural processes'. Moreover, such an approach entails, however implicitly, a requirement for continued technological advancement, and perhaps even more significantly, an assumption that ecology–technology trade-offs are both propitious and possible indefinitely. An assumption which Pierce (1992:308) considers to be ill-conceived: 'many economists have mistakenly assumed that the possibilities for substituting human-made capital for natural capital are unlimited'.

Some writers not only reject neo-liberal approaches but suggest that sustainable development can only be achieved within a socio-economic order radically different from that which exists today. O'Riordan, for example, suggests that sustainable development may well prove to be an 'inoperable concept', not least because:

17

this problem is compounded in the contemporary world by the influence of capitalist forms on the alienation of humanity from the natural world . . . it draws more from the environment than it returns yet does not pay for the loss of that environmental capital. . . . Certainly, it will mean the redistribution of wealth, technology and opportunity from the affluent to the poor in the interests of collective well-being. It will also involve patterns of development that ensure minimal resource exploitation.

(O'Riordan, 1991:7)

Redclift sees particular features of capitalism, especially uneven development, as constituting major barriers to the achievement of sustainable development:

Natural resources are systematically depleted in the accumulation drive by both private and multinational capital and the state. Ecological degradation in the South assumes emergency proportions through the mindless commitment to the economic growth strategy endemic to developed capitalism. The costs of this development are expressed not only in terms of class conflict and economic exploitation, but also in the reduction of the natural resource base upon which the poor depend for their livelihood.

(Redclift, 1987:38)

Bahro (1984) also sees the nature of capitalist production and consumption in the North as being central to contemporary environmental problems. He suggests, however, that traditional socialist prescriptions are unlikely to undermine global capitalism. Similarly, Dickens (1992:7) points out that, although modern capitalism may indeed underpin many if not all contemporary environmental problems, the idea that the solution to these problems lies in the overthrow of capitalism is, to use his terminology, 'somewhat outmoded'. This probably is the case, but a key problem here is that current political structures may well be ill-suited to the promotion of new and more sustainable modes of development. The perceived immediacy of many of the problems which sustainability encompasses creates a situation in which apparently pragmatic, but essentially ineffective, measures which address unsustainable events directly have considerable social appeal and political expediency. This situation is compounded by the inherent conservatism of extant institutions and power structures which, by their nature, tend to favour incrementalist tinkering over more radical solutions. Allied to this, an increasingly neo-liberal global political agenda has tended to promote 'cornucopianist' and market-led – race to the bottom – strategies.

Whilst the need for more radical approaches has been widely espoused (see for example, O'Riordan, 1991), these have hardly gained any political credibility, real scientific development, or for that matter widespread public support. According to Pearce (1995:10) 'translated into realisable political action, sustainable development is more about changes of emphasis than a wholesale restructuring of decision making. At best it is likely to involve a further movement of environmental concerns up the political agenda'. Thus there is an impasse: peripheral, palliative measures are possible and indeed often politically expedient, but these are at best superficial, and while it is increasingly apparent that what is really needed are more profound changes in the nature of development, the radical nature of such an agenda makes it politically and practically untenable.

Indeed the political priorities and policies of most governments seldom stray far from attempts to achieve 'sustainable economic growth'. However, as the *Caring for the Earth* report (IUCN *et al.*, 1991) points out, this is a logical impossibility – no growth can be sustained indefinitely. It is a straightforward truism that growth *per se* cannot be sustained – whatever was growing would always become infinitely large. Thus whilst many kinds of growth can be maintained in the short term none can be sustained indefinitely. This simple point has very significant implications for approaches to sustainable development predicated on the notion of sustainable economic growth, such as that formulated within the UK strategy for sustainable development (Secretary of State for the Environment, 1994). Arguments which counter this point by suggesting that economic growth is in some way qualitatively different to other kinds of growth, and thus can be sustained, are ultimately untenable. Indeed, theoretical arguments aside, history demonstrates quite clearly that the accumulation process is fundamentally crisis prone and inevitably circumscribed. In one respect this would not seem to be particularly problematic, few would argue that economic growth is an end in itself. It is, however, only a small step from this recognition to an appreciation that attempts to continuously sustain economic growth, almost inevitably involve the over-exploitation and devalorisation of both human and natural resources (Drummond and Marsden 1995; Drummond, 1996).

A new exploratory agenda

Both the theory and practice of sustainable development appear to have reached something of an impasse. This impasse reflects the congruence of several key difficulties. First, the concept is ambiguous and open to a wide variety of interpretations. Not only does this allow the idea to be (mis)appropriated to support a range of agendas, it has also caused many analysts to become unproductively bogged down in the search for the 'holy grail' of a more precise and widely acceptable definition. Second, the concept of sustainable development is not only broad but also fundamentally integrative, and no convincing methodology exists for embracing, in a single moment, the totality of the idea. Third, sustainable development is radical in that it threatens established social, political and economic structures, and linked to this, it may well be the case that established political structures are ill suited to its effective promotion. Fourth, the idea is insubstantial, in that as yet it remains insufficiently related to key areas of established theory. In particular, the political economy of sustainable development has not been fully theorised.

The actuality of the unsustainable is often cruelly simple and unambiguous, but it is now all too clear that meaningful progress towards sustainability will not emerge from strategies which simply address the concrete actuality of unsustainable events. In practice, most current approaches to sustainable development have addressed 'narrow technocratic concerns' in an atomistic and unembedded manner. The crucial point here is not simply that they have focused on specific problems and treated these discrete events, the tendency has been to dwell, quite inappropriately, on the questions of how, what, where and when, whilst a more useful starting point is why – why does the tendency to the unsustainable always seem to exist? Thus far, when this more incisive question has been asked, ensuing debates have usually remained remote from the actuality of development. To date, the political economy of sustainable development, and the significance of issues such as uneven development, remain largely unexplored.

If thinking on sustainability is to be progressed the disjunction between theory and practice needs to be bridged. If this synthesis is to be achieved, it will in itself require a much more rigorous consideration of the relevance of mainstream social theory to sustainability. Redclift begins to suggest how this theoretical clarification might be achieved:

> If we are to meet the problems presented by imminent global nemesis, we need to go beyond the assertion that such problems are themselves socially constructed. We need to embrace a realist position, while recognising and acknowledging the relativism of our values and our policy instruments. The challenge is to develop a 'third view' which enables us to assume responsibility for our actions, while exploring the need to change our underlying social commitment. We need to develop a broader and deeper foundation for the formulation of a realist policy agenda.
>
> (Redclift, 1992:22)

Although, as Redclift suggests, it would be arrogant indeed to claim any unique truth in a particular approach, what follows in this volume is a search for Redclift's third view – a conception of sustainability substantiated and empowered by social theory. The conception we attempt to develop seeks to incorporate rather than dismiss many of the other approaches which have been cited here by considering, quite literally, the ways in which a realist policy agenda might be defined and operationalised. It aims to go beyond the assertion that unsustainability is socially constructed by identifying and analysing the processes of structuration which link economic and social structures and processes to actual examples of unsustainable development.

The majority of existing approaches to sustainable development work backwards from the bottom line of biologically or morally defined sustainability metrics and thus they fail to respect either the multi-dimensional nature of sustainable development or the need for truly integrative solutions which this implies. By virtue of what they are, such approaches tend to conceptualise the situation in terms of a line one side of which lies sustainability, but beyond which lies unsustainability. The problem here is not so much that the definition of this line is often technically difficult, uncertain and contestable, although clearly it often is. The real problem lies in the fact that asking where, precisely, the line should be drawn is the wrong approach. What should be explored is why and how the line will tend to be crossed wherever it is drawn. This is the question which we have sought to address.

Approaches which begin from this position have the distinct advantage that they largely circumvent the unproductive need for any precise definition of exactly what sustainable development is. This is important because equivocation and contestation here have proved to be a major barrier to the active promotion of sustainable development. From this perspective, it is the nature of the development trajectory itself rather than the multifarious outcomes this involves which is significant. And it is the processes and mechanisms which produce unsustainable outcomes, and the conditions in which these occur, which are the focus of our concern. When sustainability is approached in this way, the need to precisely demarcate some sustainable–unsustainable boundary becomes largely redundant. Throughout this text, although we periodically refer to various practices and events as being 'unsustainable', this is merely a heuristic device. We are quite unconcerned about arguing whether or not any particular event is or is not unsustainable. We doubt that any such determination is possible, and certainly we would regard it as unnecessary and quite possibly counterproductive. Rather, our objective is to define unsustainability in dynamic and relative terms. Simply put, we generally use the term 'unsustainable' to refer to *practices or events which are more profoundly exploitative than those which preceded them* and thus, in line with our central argument, part of an 'unsustainable trajectory'. What is crucial to this approach, however, is the way in which the causality of the unsustainable is understood and explained. In the subsequent chapters of this book, we attempt to progress this understanding by drawing insights from two strands of social theory: realism and regulation theory. Both of these strands can assist in understanding the evolution of unsustainable development trajectories.

Modern conceptions of realism delve beneath surface level appearances to provide a multi-layered, powerful and practically useful basis for understanding the causality of unsustainable practices and events. The realist mode of explanation provides an interpretation of unsustainable events and practices which extends beyond that provided by more positivist interpretations, and thus one which can begin to elucidate how sustainable development might be addressed in ways which transcend the limits of what Redclift terms 'environmental managerialism'. The realist mode of explanation explains events in terms of conjunctures between structurally defined, tendentially expressed causal mechanisms and contingent factors. Most current approaches to sustainability attempt to influence the causality of unsustainable events at the level of contingency. This is clearly inadequate, but realism provides the opportunity to explore the potential to influence the tendencies involved as well as contingent factors. Moreover, it provides the opportunity to do this in an objective and specific manner. Realism suggests that actual events depend not just on contingent factors and whether particular mechanisms are present, but also on whether these mechanisms are 'activated'. A central tenet of our argument is that mechanisms significant in the causality of unsustainable modes of development are selectively legitimated and empowered by what regulationists term the 'mode of social regulation', and that this selectivity is biased. Because of their particular bias, current modes of social regulation tend to condition development in unsustainable directions. Understanding this conditioning so as to be able to modify it in an objective way may well be crucially significant to the achievement of sustainable development.

Insights from regulation theory, which is centrally concerned with the contradictions and crises which emerge within capitalist economies, can substantiate this realist conception of the causes of unsustainability. Implicitly at least, regulation theory is centrally concerned with why and how some aspects of development are sustained whilst others are devalued and degraded. If we accept that sustainability will, necessarily, be sought and achieved within capitalist economies, insights from regulation theory have some considerable relevance to sustainability debates. We argue that an extension of regulationist thinking can throw light on why some unsustainable events and practices come about. From this perspective disequilibria are generated from within the capitalist accumulation system and are exported beyond the system itself. Although a regulationist perspective suggests a limited competence for human agency, we cannot, for example, simply construct new modes of social regulation. We argue that if we understand more specifically which elements of the mode of social regulation are involved in conditioning development in unsustainable ways, it may well be possible to condition a more sustainable future through positive intervention. This point will be reassessed in the conclusion.

If thinking on sustainability is to be progressed, attention must be directed to the ways and potentialities of present systems of economic and social regulation, assessing how these may begin to bring about institutional and value change at the social and economic level. Potentially effective policy formation requires a thorough understanding of how the dynamic and volatile nature of development itself predicates the adoption of practices which involve unsustainable forms of exploitation and how these are able to achieve their own social and political legitimacy. As Benton (1994:50) suggests, we need to recognise that what is 'out of control' is not some mysterious *telos* of history, but the key institutional process of corporate control, state power and scientific innovation.

A realist methodology allied to an appreciation of the role of modes of social regulation in capitalist societies provides a conceptual framework within which progress can be made here. Historically, 'regulation' has been centrally concerned with maintaining the value of capital and fixed assets. The achievement of sustainable development will obligate a broader remit in the future. If this expansion is to be promoted, an understanding of what must be regulated and how and at what level this might best be achieved becomes crucial.

2

REALISM

Questions of ontology – understanding the causality of the unsustainable

Introduction

Beginning with a review of modern conceptions of realism, this chapter discusses how these define an ontology within which thinking about sustainability can be progressed. Realism provides a multi-level conception of causality, and one which, we contend, is a potentially powerful basis for understanding the nature of unsustainability in contemporary society. From a realist perspective, events are understood as outcomes which reflect both tendentially expressed, structurally defined mechanisms and contingent conditions. We argue that a particular significance attaches to the conditions which selectively activate key causal mechanisms, not least because these potentially form a basis for new strategies for the promotion of sustainable development. The final sections of the chapter consider first the general methodological considerations posed by a realist approach, and subsequently to the actual research methods used during the empirical research discussed later in this volume.

Realism and sustainable development: the problem of causality

If we accept that measures which address what are perceived to be unsustainable events in a direct and unembedded manner are not, in themselves, sufficient to ensure sustainable development, one corollary of this is that more effective action depends on the development of policies based on a deeper understanding of why and how the unsustainable comes about. From this perspective, the need is to focus on understanding what it is that causes unsustainable events and practices. Developed largely from the work of Bhaskar (1975, 1979), modern conceptions of realism posit a particular understanding of causality which is of considerable use here.

Critical realism provides a means of 'combining insights from a number of disciplinary perspectives without sinking into deep and irretrievable eclecticism' (Dickens, 1992:177), but its particular utility here lies in the opportunity which it affords to move beyond atomistic approaches to sustainability. The nuanced and multi-layered explanation of causality provided by realism can elucidate new and potentially more useful bases for regulation. As Lovering (1990:39) puts it:

> Critical realism holds that reality, including society, is made up of deep structures which condition and make possible the 'events' we observe in everyday experience and, importantly, in scientific research.

We are concerned here to explore how development is 'conditioned' in ways which make unsustainable practices and events the norm, and to consider whether and how it may be possible to modify this conditioning to produce more sustainable patterns of development. In exploring the potential of such an approach it soon becomes clear that neither general assertions regarding the exploitative and structural nature of capitalism nor vague and unembedded prescriptions for value change in society are adequate. The relationship between the abstract and the concrete, and in particular the processes of structuration which link the real and the actual need to be explored in ways which make it possible to identify purposive, specific and potentially effective forms of intervention. Understanding sustainable development in these terms is problematic, but it is, we contend, the key to a sustainable future.

The realist mode of explanation

According to Outhwaite (1987:19) realism is a 'common-sense ontology, in the sense that it takes seriously the existence of things, structures and mechanisms revealed by the sciences at different levels of reality'. Bhaskar (1975) identifies three such levels or domains (see table 2.1).

Objects within the domain of the real give rise to structures, which by their nature produce certain tendencies or mechanisms which act as causal agents, but such tendencies are invariably mediated through contingent

Table 2.1 The three domains of reality

	Domain of the real	Domain of the actual	Domain of the empirical
Mechanisms	✓		
Events	✓	✓	
Experiences	✓	✓	✓

Source: Bhaskar 1975

conditions. Thus a particular mechanism may or may not produce a particular event. As Sayer (1984:99) suggests 'for any particular set of conditions, the results occur necessarily by virtue of the nature of the objects involved, but it is contingent which conditions are present'. Moreover, particular events within the domain of the actual may well be subject to complex patterns of causation involving plural and possibly countervalent causal mechanisms (see figure 2.1). This is a two stage argument: first, mechanisms may not be activated and, second, where they are, the effects depend on conditions. As Sayer suggests:

> Events are causally explained by retroducing and confirming the existence of mechanisms, and in turn the existence of mechanisms is explained by reference to the structure and constitution of the objects which possess them. Where the same events are co-determined by several distinct causes, they may also be explained by calculating the relative contributions of each mechanism.
>
> (Sayer, 1984:214)

The realist mode of explanation and its relevance to the promotion of sustainable development is perhaps most easily understood by means of an allegorical example. Teachers faced with the problem of assessing mixed ability groups of students often set questions which they formulate to allow 'differentiation by outcome'. That is to say questions which can be answered in different ways – some of which it is assumed are more sophisticated than others – but all of which are apparently sensible and correct. A good example of this is the seemingly straightforward question 'why do rivers meander?'. Secondary school pupils can, and do, answer this question in terms of differential patterns of erosion and deposition as a river flows more quickly around the outside of a bend than it does on the inside. Equally, however, faced with the same question undergraduate fluvial geomorphologists might well attempt to answer by reference to the laws of thermodynamics and the concept of entropy. In these terms a river meanders because it must seek an equal distribution of free energy throughout its length.

These examples of how we might seek to answer the question 'why' an event occurs – i.e. to explain its cause – are useful in that they help demonstrate the realist mode of explanation. Implicitly, at least, the second type of explanation mirrors a realist approach to the understanding of causality. To understand why this is so we must consider the question in a little greater detail. In the second answer, the explanation was formulated in terms of a generally applicable if unobservable determinant of causation – the laws of thermodynamics. However, whilst all rivers are subject to this same structurally determined causal power – the laws of thermodynamics ought to apply to all rivers – empirical observation would quickly suggest that not all rivers meander in practice. This apparent lack of general applicability might be

The structure of causal explanation

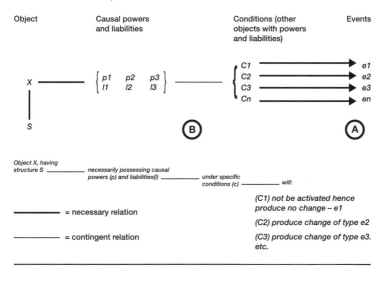

**Object X, having
structure S** ──────── *necessarily possessing causal
powers (p) and liabilities(l)* ──────── *under specific
conditions (c)* ──────── *will:*

──────── = necessary relation

──────── = contingent relation

*(C1) not be activated hence
produce no change – e1*

(C2) produce change of type e2

*(C3) produce change of type e3.
etc.*

Structures, Mechanisms and Events

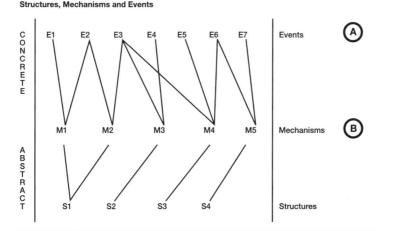

Policies to promote sustainable development normally target intervention at level **A** and seek to address unsustainable events by modifying contingent conditions. Intervention at level **B** is fundamentally integrative and has the potential to prevent rather than mitigate the outcomes produced. From a realist perspective, causal powers and liabilities are articulated through tendentially expressed mechanisms. In itself the presence of a mechanism is not sufficient to cause an event to occur. Actual events depend on both contingent conditions and whether or not the mechanism is activated. Effective intervention at this level of causality requires knowledge and understanding of what mechanisms are significant and the specific institutional and social conditions which activate them.

Figure 2.1 Realism and sustainable development

Source: after Sayer, 1984

seen as an indictment of the realist explanation. In fact, however, this inconsistency encompasses the essence and strength of the realist approach. Indeed, much of the appeal of the realist ontology lies in its ability to incorporate heterogeneous and differentiated outcomes within a unified multi-level mode of explanation.

Realism does not suggest that an object in the domain of the real – in the case of the meander, the laws of thermodynamics – acts directly to cause an event. Rather causality is understood in terms of tendencies which may or may not be realised in practice. Thus whilst a river will always tend to meander it does not always do so. The tendency to meander may be necessary for a meander to occur but it is not, in itself, sufficient for this to happen. An event – a particular meander in a particular place at a particular time – will only occur if certain contingent conditions are met. In this case, relevant contingencies might well include geological and climatic factors. Equally, the historical pattern of development in a particular case may constitute a significant element of contingency. In the case of the meandering river, for example, the concrete reality of the present must reflect historical patterns of erosion, and thus it must also effect the level of free energy in the system and it follows the potential for erosion which now exists. Thus, in practice, some rivers meander while others do not. Faced with the same tendency to achieve a redistribution of free energy, the river may, for example, become braided rather than meandering. In which case the tendency is still expressed in the domain of the actual – braiding will also serve to redistribute energy. However, whilst realist theoreticians speak of 'conjunctures' of real causal mechanisms and contingent factors which combine to produce a particular event, it is important not to conceptualise structurally defined tendencies as often latent mechanisms which are, on occasion, triggered or enabled by particular contingent factors. A realist philosophy rejects the idea that objects in different domains are discrete and understandable in isolation. Rather it stresses the mutuality and interdependence of objects in different domains even where outcomes may be different. Thus whilst a tendency to equalise the distribution of energy within the system may well be significant in meander formation, the formation of a meander will itself reconfigurate the real object and its associated tendencies. This must be so, for the formation of the meander necessarily redistributes energy in the river system.

To understand why this kind of explanation is relevant to thinking about sustainable development, it is useful to develop this analogy a little further and consider what type of strategy might be used to prevent a river from meandering. One approach would be to devise a direct and literally concrete engineering solution to prevent erosion along the stretch of river in question. Such a solution is closely analogous to attempts to define and enforce sustainability limits – maximum sustainable yields, pollution controls, designated areas or whatever. A more subtle approach, and quite probably a more effective one, would be to build a weir or to reforest the catchment to modify the hydrological processes occurring and thus regulate the river's flow regime. A strategy such as this would influence the tendency to meander rather than prevent its expression more directly. The contention here is that it may be possible to address unsustainable events in a manner analogous to this, and that a multi-level conception of causality posited by a realist approach potentially provides a powerful conceptual framework within which to pursue such an agenda.

Consider, for example, the environmental problems arising from the intensification of agriculture, see figure 2.2. Clearly there are a number of such problems and these along with a number of social problems relating, for example, to intergenerational transfer, farm debt or age structures in farming, might well be categorised as examples of unsustainability. In practice, various measures have been used to obviate these problems including pollution controls, area designations and grants to support apparently desirable practices such as reafforestation. These responses, as with most planning for sustainability, are essentially reactive and outcome based. A more realist approach would also consider and address the structures and tendencies which condition the nature of modern agriculture.

It is clear enough that the vast majority of contemporary agricultural practice is, in large part at least, conditioned by a range of factors including: the nature of national and multinational food systems, price support and institutional and corporate support for technological development. It is also

- The human health effects of pesticide and fertiliser residues, heavy metals, feed supplements and other contaminants in soil, water bodies, food products and the food chain;
- The diminution and partition of biotopes valued for nature conservation;
- The contamination of ground and surface waters and the eutrophication of surface waters by nitrates and phosphates leading to local health risks, decline in the quality of aquatic resources, decline in recreation values and increased water supply costs;
- Agricultural pollution problems associated with the growth of intensive animal husbandry;
- Air pollution from intensive animal production, manure spreading and crop spraying;
- The salinisation of soils, which is contaminating water supplies and causing losses in soil productivity and landscape amenity values;
- Losses in landscape amenity and wildlife habitat caused by the amalgamation of farms, the growing emergence of monocultures, the removal of hedges, walls and terraces, the draining of wetlands and the deterioration and destruction of traditional farm building; and
- Soil compaction, erosion and pollution which have led to productivity losses, declines in the quality of water resources and reduction in the capacity of water storages.

Figure 2.2 Environmental problems arising from the intensification of agriculture

clear that as the modernisation process progressed it has come to involve increasingly exploitative practices, and that this trend is continuing, not least through the promotion of new technologies such as genetic engineering. Although we do not want to focus too deeply on identifying and substantiating exactly which mechanisms have been significant in defining the trajectory of agricultural development at this juncture, three points are worthy of further clarification. First, it is unproductive and indeed, we believe, ill-considered to try to establish, in some 'scientific' manner specifically which of the outcomes detailed in figure 2.2 are actually unsustainable. Better and more useful to focus on the general and unsustainable dynamic of development in the agricultural sector. Second, and related to this, if we accept that the important issue here is the unsustainability of the trajectory of agricultural development rather than the specificity of which this is constituted, it is then quite clear that the potential of specific measures which seek to address each and every apparently unsustainable outcome in an atomistic and unembedded way is very limited. Third, and following from this, the need is to consider how, and indeed specifically how, this trajectory is conditioned and how this conditioning might be modified. One way to do this is to adopt a realist approach and focus on identifying and understanding precisely which structures and mechanisms are involved in the causation of the unsustainable, and importantly, which conditions are significant in selectively empowering these mechanisms. This does not, necessarily, imply a grand design approach to sustainability. As Sayer (1984) suggests, it is ill-conceived to conflate 'structural' and 'macro' – structural does not necessarily imply large scale and general, anything and everything which is real has structure. Thus until specific mechanisms have been identified as being significant in the causation of the unsustainable, it remains to be seen at what scale appropriate intervention would focus. Whatever the scale, however, we believe that it would be logical and potentially more productive to seek to regulate the mechanisms involved rather than the actual events which they predicate, whether these are soil erosion, forest depletion or any other kind of unsustainable outcome.

Realism and the promotion of sustainable development

The principal utility of realist analysis to our agenda lies in the particular understanding of causality it provides. This can potentially allow us to move beyond approaches to sustainable development which address the direct causes of apparently unsustainable practices and events and to formulate new and different approaches based around deeper and more dynamic conceptions of what is and is not sustainable. The multi-layered mode of explanation provided by realism posits a view of causality which encompasses both structurally defined, tendentially expressed causal mechanisms, and contingent factors. This deeper perspective is significant because we are concerned to

30

broaden the scope of sustainability policy beyond measures which address what are effectively contingent factors, and to explore the possibility of regulating the tendencies involved rather than the specific outcomes which they underpin.

Crucially, the realist mode of explanation suggests that relevant mechanisms need to be 'activated' before they become significant causal factors. Understanding not just which mechanisms are involved, but also why and how these are selectively legitimated and made consequential is the key to understanding why present day modes of development tend to involve unsustainable outcomes; it may also be the key to formulating new types of sustainability policy. The realist mode of explanation suggests that the mechanisms which link objects and structures to actual events are variously empowered or rendered inconsequential. This process of 'activation' effectively determines the nature of the events which are actually realised. We will argue that this process is currently biased in ways which condition development in unsustainable ways. Understanding the processes of structuration through which particular mechanisms are legitimated and empowered is thus crucially important to understanding why and how development tends to the unsustainable. From this perspective, the institutional and social conditions within which causal mechanisms are expressed are highly significant because the structuration embodied in these conditions is a fundamental influence on the nature of development. Within this structuration, it may be possible to find both new objects of regulation and new forms of regulation which will be useful in the promotion of an objective such as sustainable development. In this sense realism may provide a new and useful understanding of why and how development tends to the unsustainable and how this might be changed.

For example, we might contend that the mechanisation of agriculture is a mechanism which has been closely implicated in a number of the environmentally degrading outcomes described in figure 2.2. To some extent, levels of mechanisation and indeed the environmental impacts of this process will depend on purely contingent factors – a particular farm may be topographically unsuited to modern farm machinery, or the local ecology may be particularly resilient. However, a range of other conditions also affect the adoption of new technologies – the financial resources of the farmer concerned, his or her access to credit and hence the nature of the financial system, the conditions of production and the regulation of the food system, or for that matter the individual farmer's attitude to farming and to the environment. The point here being, that intervention at the level where general tendencies are expressed through a particular mechanism may well be sufficient to prevent the outcomes associated with that mechanism from being realised in practice. One problem, however, is that if it is to be effective, policy in this area must move beyond the vague and unembedded assertions which pervade some sections of the sustainability literature, to identify specifically and substantially which mechanisms, and which conditions, are implicated in the definition of unsustainable development trajectories.

Two further potential problems with the approach being developed here involve, first, the limited potential which the researcher has to understand the real and, second, the limited potential of human agency to translate any such understanding into strategy. By rejecting the 'correspondence' version of truth postulated by more positivist epistemologies (Keat and Urry, 1982:18), critical realism accepts that:

> All knowledge must be considered to be not only fallible but also necessarily open to immanent, or ongoing, critique. In short 'Truth' must be considered to be conditional and not as absolute.
>
> (Pratt, 1995:66)

This acceptance holds both significant methodological connotations and implications for the utility of the realist approach itself (Keat and Urry, 1982:40). Bhaskar's (1979) 'transcendental' view of realism postulates a reality external to social construction, and thus essentially beyond the reach of the researcher. However, this does not necessarily mean that a 'practically adequate' model of causality cannot be constructed (Sayer, 1984; Latour, 1988). As Sayer (1984:330) suggests:

> We can't get outside discourse to see how it compares with real objects, but it is evident from observation and action within a particular world-view or discourse that some conventions about what is the case hold and others don't.

The suggestion here then is that whilst a researcher can only exist in the domain of the empirical and can thus have no or only limited objective knowledge of the real, they can nevertheless develop an understanding of the real which to most intents and purposes holds good, and which may be adequate to inform policy. However, as Lovering points out, a realist understanding predicates a view of history within which human agency is both circumscribed and uncertain of its potential efficacy:

> The picture which critical realism offers is one in which individuals enter into a world which is not of their own choosing, and once there they act in ways which partly reproduce, partly transform the structure of that world. But their understanding and ability to control these structural effects are severely limited, and social entities and structures are often reproduced as unintended effects of individual actions.
>
> (Lovering, 1990:38)

Although Lovering's caveats are clearly well founded and significant, it may well be that a realist ontology can still serve to inform and thus empower some agency over others. Indeed, in many respects the agenda being explored here is concerned to understand just how we can use our 'limited abilities' to maximum effect.

Realist methodology

Although it is possible to make a convincing case for a realist approach to understanding sustainability issues in terms of its theoretical legitimacy and potential utility, the application of such an approach is less than straight-forward in practice. As Pratt points out:

> On reflection it can be noted that the appropriation of critical realism by geographers has been at best partial. For a perspective that stresses the integral importance of empirical work it is a supreme irony that the complementary, practical, element is almost totally under-developed.
>
> (Pratt, 1995:67).

In many ways it is the very nature of the realist mode of explanation which makes realist research difficult to conduct in practice:

> In explaining any particular phenomenon, we must not only make reference to those events which initiate the process of change: we must also give a description of that process itself. To do this, we need knowledge of the underlying mechanisms and structures that are present, and of the manner in which they generate or produce the phenomenon we are trying to explain. In describing these mechanisms and structures we will often, in effect, be characterising the 'nature', 'essence', or 'inner constitution' of various types of entity.
>
> (Keat and Urry, 1982:30)

The ontological basis of realism provides a multi-layered but unified mode of explanation and the task for realist researchers is to embrace, in a single moment, the full extent of the realist view. In practice, the problem becomes one of approaching research in ways which are necessarily partial but which still respect the unity of the realist mode of explanation:

> To be practically adequate, knowledge must grasp the differen-tiations of the world; we need a way of individuating objects, their attributes and relationships. To be adequate for a specific purpose it must abstract from specific conditions, excluding those which have no significant effect in order to focus on those which do. Even where

we are interested in wholes we must select and abstract their constituents.

(Sayer, 1984:80)

According to Cloke *et al.* (1991:148) 'the practice of realism involves two basic requirements: (1) theoretical categories, so as to "get at" necessary relations; and (2) empirical study, so as to "get at" contingent relations'. Sayer advocates a process of 'synthesis' which he explains in these terms:

> Abstract theoretical research deals with the constitution and possible ways of acting of social objects and actual events are only dealt with as possible outcomes. Examples include theories of value in economics and those theories of social class which define class in terms of internal relations. Concrete research studies actual events and objects as 'unities of diverse determinations', each of which have been isolated and examined through abstract research. By contrast, the method of generalisation tends not to involve abstraction, at least not self-consciously and treats objects as simple rather than concrete. Its main purpose is to seek regularities and common properties at this level. We might also add a fourth type – 'synthesis'; that is, research which attempts to explain major parts of whole systems by combining abstract and concrete research findings with generalisations covering a wide range of constitutive structures, mechanisms and events.
>
> (Sayer, 1984:215)

Outhwaite (1987:58) advocates an approach to critical realist research which focuses on the postulation of possible causal mechanisms the validity and significance of which can then be analysed; first by collecting evidence for their existence and subsequently by eliminating possible alternatives. An important point here is the particular meaning ceded to the idea of a mechanism in realist thinking:

> In asking about the structure generating some power of some entity, we are asking about a 'mechanism' generating an 'event'. A mechanism in this sense is not necessarily mechanical in the sense of Newtonian mechanics. It could be an animal instinct, an economic tendency, a syntactic structure, a Freudian 'defence mechanism'.
>
> (Collier, 1994:43)

A major difficulty here lies in the fact that it is not obvious how one might progress directly from empirical observations of concrete events to the identification and analysis of key causal mechanisms in the domain of the real. Realist methodology confronts this problem directly by suggesting that

34

when 'real objects' are unobservable, they can still be identified through a process of abstraction in which the 'necessary' or 'internal' relations between objects are analysed. Such relationships are often explained by the analogy of a tenant – for there to be a housing tenant there must be a landlord, and it follows that if there is a landlord property rights must exist, and so on (Allen, 1983). Pratt (1995) uses the term 'retroduction' to describe the process through which causal mechanisms are identified, corroborated and subsequently substantiated.

Sayer explains the significance of this type of relationship more fully:

> In making abstractions it is helpful to distinguish between relations of different types. The term 'relation' is a very flexible one but there are some significant contrasts implicit in its various uses. A simple distinction can be made between *'substantial'* relations of connection and interaction and *'formal'* relations of similarity or dissimilarity. . . . Clearly things which are connected need not be similar and vice versa. . . . Another useful distinction can be made between *external, or contingent relations* and *internal* or *necessary relations.* The relation between yourself and a lump of earth is external in the sense that either object can exist without the other. It is neither necessary nor impossible that they stand in any particular relation; in other words it is contingent. . . . By contrast, the relation between a master and slave is internal or necessary, in that what the object is is dependent on its relation to the other; a person cannot be a slave without a master and vice versa. Another example of the relation is the relation of a landlord and a tenant; the existence of one necessarily presupposes the existence of the other.
>
> (Sayer, 1984:88)

In practice, most actual conjunctures are more complex and less straightforward than the idealised exemplars often used to demonstrate realist methodology would suggest. Moreover, the problem is not simply one of sifting out spurious contingent factors. Plural and possibly counteractive mechanisms may well be involved. In practice, this means that once mechanisms and hence structures have been tentatively identified, their nature and relevance have to be verified and substantiated through a reflexive exploration of their characteristics and properties and their correspondence with both abstract notions of structural properties and actual events. Thus a key element of the realist methodology, and in practice one which differentiates it from positivist methods, is the emphasis it places on moving iteratively to a more substantial and convincing model through a process of conceptualisation and reconceptualisation in which the researcher's understanding of the constitution and significance of mechanisms is progressively revised. As Pratt (1995:67) explains:

> The process may not be a direct or linear one, often it is an iterative one, the model being refined in an ongoing process. Whilst this may seem a trivial point it does severely challenge existing modes of scientific endeavour, both implying a far more exploratory structure and a challenge to the common form of presentation of results ... the process of conceptualisation and reconceptualisation is central throughout the whole endeavour.

Moreover, it is highly unlikely that significant aspects of causation, be they structural or contingent, can be isolated from the unpacking of any single event. At best, in such a case it is only possible to suggest that several elements are present and that one or more of them may be significant. Exactly this question of selectivity has been a major problem in the study and practice of sustainable development. That is, how does one arrive at an effective prioritisation of key mechanisms at different levels of power and influence in order to progress towards a more sustainable end? In practice, if we wish to isolate significant elements of causation it is probably necessary to analyse a range of related events. Whilst such a methodology is vulnerable to the problem of the method of difference, that is to say what is common may not be causally relevant, and vice versa, and thus cannot be totally conclusive, it may prove sufficient in this context.

A further problem here is that the process of retroduction, or at least the way in which it is practised, is always likely to be influenced by pre-existing notions concerning the nature and significance of the structures being considered. For example, if capitalism is pre-emptively understood to involve tendencies to disequilibrium and crisis (as in regulation theory), the research is always likely to find evidence to corroborate mechanisms which would reflect this interpretation within the reflexive process of substantiation and refinement. Thus, in practice, analysis may well still reflect the bias of the researcher. Ideally, however, properly conducted realist research should inform both the theoretical categories involved in the process of abstraction as well as explanations of why particular events come about.

An additional complication for the realist researcher lies in the fact that events are not seen as simply reflecting conjunctures between mechanisms and contingent factors. The ways in which mechanisms are selectively 'activated' by the conditions in which they occur is also crucially significant. As Lovering (1990:32) points out, connections in any specific historical instance are not only likely to be complicated, they must also be understood theoretically as 'mediated effects'. For example, it can be argued that in capitalist societies, the logic of capitalist dynamics is mediated through particular and different modes of social regulation. Thus actual events are richer, more diverse and less strictly determined than any purely fundamentalist conception could allow. Because of this, the processes through which significant causal mechanisms are identified and substantiated cannot simply evaluate the correspondence of mechanisms to the perceived structural properties and actual events. The evaluation must relate to the particular context in which a mechanism occurs. Certainly, this context is held to be highly significant in this thesis – not least because,

> it should be noted that critical theory does not simply *replace* research on what *is* with criticism of what is, plus assessments of what *might be* from the point of view of emancipation. It would be a poor critical social science which imagined that it could dispense with abstract and concrete knowledge of what is in society. If certain mechanisms are to be overridden or undermined and new ones established we need abstract knowledge of the structures of social relations and material conditions by virtue of which the mechanisms exist . . . we can also see that it would be poor abstract or concrete research which was unaware of the fact that what *is* need not necessarily be, and which failed to note that people have powers which remain unactivated in the society in question but which *could be* activated.
>
> (Sayer, 1984:256)

Realist research techniques

The question of precisely which research techniques are most appropriate to a critical realist research agenda has been a matter of some debate (Sayer, 1984; Fielding and Fielding, 1986; Dale *et al.*, 1988; Burrows, 1989), and as Pratt (1995:67) suggests, the methodology for 'putting critical realism to work' remains, 'a rather vague recipe book'. The central question which realist researchers need to consider is: 'What, in practice, is so different about research informed by critical realism?' (Pratt, 1995:67). When this question is asked, it soon becomes clear enough that the nature of the realist mode of explanation means that techniques considered to be appropriate within alternative epistemologies are often inadequate for realist analysis. For example, methods which seek to establish and verify, for example through empirical invariance, a direct causal relationship between factor x and event y are inappropriate because they assume that the cause of y can be adequately explained in terms of x (Bhaskar, 1994:19). Similarly, an understanding of the intuitively determined properties of particular structures may form a key element of any realist analysis, but this alone cannot provide an adequate explanation of events within this perspective. For example, 'In the world according to [Marxist] "fundamentalism", the fact that the economy is *capitalist* is of overriding significance . . . the task of analysis is to draw out the connections between observable development and the underlying dynamics of capitalist class relations' (Lovering, 1990:32), but this is discordant with a realist approach because it posits a direct and ultimately deterministic and teleological mode of explanation. Such logic may be valid, but it is again partial. The problem is not so much that these techniques are incompatible with a realist mode of enquiry. Rather they are insufficient to produce the holistic explanation essential to realist analysis.

Realist research does not reject research techniques used in other epistemologies, rather it seeks to place them within a broader model (see table 2.2). Pratt (1995:68) writes of attempts to 'link' or 'combine' different approaches. However, it is clear that realist methodology is necessarily more than an eclectic combination of techniques. For example, it is not concerned to validate results through finding empirical invariance. Rather the realist researcher is concerned to establish, albeit through an iterative process of repeated reconceptualisation, a convincing and 'practically adequate' model which captures, unifies and elucidates the realist mode of explanation (Pratt, 1995:66; Sayer, 1984:66).

The objective of realist research is to uncover significant causal powers, and this objective prescribes the use of investigation techniques which are

Table 2.2 Realism and sustainability: an intensive research approach

Type of research question	How does an economic sector operate in a particular case or small number of cases? What unsustainable outcomes are produced in that particular sector? What did agents operating in that sector actually do?
Types of groups and individuals studied	Causal groups (practitioners, key decision-makers, stakeholders, partners, firms, farmers, etc. in the sector)
Type of explanatory account produced	Causal explanation of how unsustainable outcomes were produced in a particular context or set of contexts, though not necessarily representative ones
Typical methods	Study of individual agents and institutions in their causal contexts, interactive and face-to-face interviews, historical analysis, documentary evidence, ethnography, on-site visits, participant-observation and qualitative analysis
Type of generalisation	Actual patterns of unsustainability are unlikely to be representative, average or generalisable. Underlying causal mechanisms may be generalisable so long as their basic structures are present in other contexts, e.g. causal powers and liabilities of sector actors may be generalisable across contexts, although the conditions of activation will vary
Appropriate tests	Corroboration. Usually involves comparative case studies to check whether conditions in different contexts activate the same causal mechanisms. If mechanisms are not the same, comparisons isolate different conditions
Policy focus	Intervention directed at preventing unsustainable outcomes by changing the conditions in which causal mechanisms operate

Source: after Sayer 1984:222

less structured than would be the case within positivist epistemologies. The type of information required for a realist analysis is best achieved through informal techniques which maximise information flows by allowing respondents to highlight the significance of their own powers through their own, albeit subjective, interpretations of causal processes. The logic of this is that realist research in the social sciences should utilise unstructured interviews and open questions and be flexible enough to respond to the direction and emphasis provided by the respondents (Pratt, 1995:69). Accordingly, the realist researcher is concerned to identify and investigate, not a representative sample of the population, but rather those agents with significant causal powers (to identify the inherent properties of the objects involved and the ways in which these properties relate to each other to produce particular outcomes). The individuals upon which the research focuses

> need not be typical and they may be selected one by one as the research proceeds and as an understanding of the membership of a *causal* group is built up.
>
> (Sayer, 1984:221, emphasis in original)

3

REGULATION AND THE CONDITIONING OF THE UNSUSTAINABLE

Questions of theory and practice

Introduction

This chapter begins by outlining the main tenets of regulation theory. Consideration is then given to the ways in which modes of social regulation condition the nature of development. In particular, it is suggested that current modes of social regulation selectively legitimate and empower strategies which sustain extant social formations by translating the contradictions which emerge within these into materially and morally significant forms of unsustainability. According to regulation theory, modes of social regulation are constituted in the institutions, structures, norms and values which cede coherence to particular phases of capitalist development. In effect, particular modes of social regulation define many of the conditions relevant in the activation of various causal mechanisms – many of which are relevant to sustainability. These institutions, norms of behaviour, values, etc., represent the canalisation of history, socially constructed channels between the real and the actual through which the currents of development are regulated and within which the general nature of specific events is conditioned. By defining rights, constraints and powers which in turn influence the ways in which real causal mechanisms are expressed in practice they serve to license and to some extent direct the nature of development. We argue that understanding the role of regulation in capitalist societies in this way can inform the ways in which we conceptualise the idea of sustainable development and the ways we seek to achieve it. The final sections of the chapter outline the techniques used to apply this research agenda in practice.

A theory of regulation: capitalism as a non-equilibrating process

In chapter 2, it was argued that a realist ontology may well be appropriate

and useful to thinking on sustainability. Regulation theory is itself founded
in a realist ontology, as Jessop explains:

> The Marxian ontology implies that the real world is a world of
> contingently realised natural necessities. This world is triply com-
> plex: it is divided into different domains, each having its own causal
> powers and liabilities; these domains are involved in tangled hier-
> archies, with some domains emergent from others but reflecting back
> on them; and each domain is itself stratified. Comprising not only a
> level of real causal mechanisms and liabilities but also the levels on
> which such powers are actualized and/or can be empirically exam-
> ined. For Marx the causal powers and liabilities in the domain of
> social relations were typically analysed in terms of tendencies and
> counter-tendencies which together constitute 'laws of motion'. These
> 'laws' operate as tendential causal mechanisms whose outcome
> depends on specific conditions as well as on the contingent inter-
> action among tendencies and counter-tendencies; thus, in addition to
> *real* mechanisms, Marx also described their *actual* result in specific
> conjunctures and sometimes gave empirical indicators for these
> results.
>
> (Jessop, 1990:162)

Regulation theory has been developed, mainly but not exclusively in
France, building on the work of Michael Aglietta with the original rationale
for the project stemming directly from the recognition that capitalism is not
an equilibrating process (Aglietta, 1979:10). Thus, in so much as the logic
of sustainability implies the need for some kind of equilibrium, insights
from regulation theory may have considerable relevance to sustainability
debates (Pierce, 1992; Drummond and Marsden, 1995a). Within this, a central concern with the ways in which contradiction and crisis emerge and are
subsequently averted or at least postponed through modes of social regulation may well inform our thinking as to why and how unsustainable events
come about and how they might be avoided.

The understanding of capitalist development over broad time horizons has
been the major objective of regulation theory to date. According to Aglietta:

> Economists confronted with the transformations and crises of con-
> temporary Western societies, and with the troubling future of the
> capitalist system as a whole, can find no foothold in general equi-
> librium theory. To take refuge in partial investigations, half
> empirical, half theoretical, only compounds the confusion. The way
> forward does not lie in an attempt to give a better reply to the
> theoretical questions raised by the orthodox theory, but rather in an
> ability to pose quite different questions. This means a collective

effort to develop a theory of the regulation of capitalism which isolates the conditions, rhythms and forms of its social transformations ... The term 'regulation', whose concept it is the task of theory to construct, denotes the need for an analysis encompassing the economic system as a whole. This analysis should produce general laws that are socially determinate, precisely specifying the historical conditions of their validity.

(Aglietta, 1979:15)

As Clarke explains:

For Aglietta the market is not an autonomous mechanism of the hidden hand, but a social institution, whose regulatory function cannot be presupposed. The operation of the market has to be conceived within the framework of a theory of regulation which establishes the possibility and limits of social and economic reproduction through an analysis of the complex web of historically specific and socially determined modes of regulation.

(Clarke, 1988:62)

In particular, regulation theory has attempted to explain how capitalism could survive despite crises congenital to the logic of capital accumulation. As Moulaert and Swyngedouw put it, the regulationist approach is concerned to theorize:

(1) the social and economic forms that channel the contradictions resulting from previous phases of sustained accumulation up to the moment that a major crisis arises, and (2) the development of new socio-economic forms that result from the crisis and the actions taken by (groups of) social agents. Embedded within this approach is the possibility of different forms of crisis: (a) short 'conjunctural' crises requiring minor adjustments (for instance, incremental technological changes, expanding spatial divisions of labour, and institutional adjustments); (b) structural crises (or crises of a particular mode of development) leading to qualitative changes in the organisation of the accumulation process; (c) crises resulting from fundamental contradictions in the capitalist mode of production itself.

(Moulaert and Swyngedouw, 1989:329)

Thus far, little of this literature has incorporated sustainability issues, and conversely, the unsustainable has generally been interpreted in terms of what Moulaert and Swyngedouw define as 'conjunctural crises'. That is, examples of unsustainability have been understood as discrete, unembedded events which can be satisfactorily addressed in a direct manner. This may be inappropriate. It may well be that the unsustainable is often more properly and more usefully understood and addressed as outcomes which reflect the contradictions inherent in the second and third types of crisis outlined above.

By rejecting the notion that the conditions needed for the functioning and progression of capitalism are created in some miraculous way as structures reproduce themselves quasi-automatically, regulation theory cedes a certain, albeit limited, consequence to human agency. The suggestion is that conflict is regulated, avoided or at least postponed, through an ensemble of norms, institutions, organisational forms, social networks, and patterns of conduct, which sustain the conditions necessary for continued capital accumulation. Thus regulation theory replaces the notion of 'reproduction' with one of 'regulation'. Regulation is, however, inevitably imperfect and any regime of accumulation will always tend to be crisis prone and temporary. Our argument is that because 'regulation', in this sense, is primarily concerned with averting crises inherent in the capitalist dynamic, indeed it achieves its own validity through its potential to maintain the established regime, it inevitably tends to involve increasingly profound forms of exploitation. Thus it tends to condition development in ways which mean that unsustainable outcomes become the norm. This sort of crisis can be addressed in different ways, some of which may appear at least to be benign involving, for example, expanded consumption or the application of new technologies. It is a moot point as to whether or not these fixes are, by their nature, always increasingly exploitative – consider the role and effects of colonialism or the potential for environmental degradation inherent in many new technologies. It is quite clear, however, that in the end, effective elements of regulation will always need to become increasingly exploitative. Effective regulation, apparently effective fixes, are necessarily only temporary and the point is that subsequent fixes need to address even more profound crises and thus will always tend to involve what by any definition are unsustainable outcomes.

Key concepts

Although the contradictory nature of capitalist accumulation is such that it is inevitably crisis ridden and temporary, regulation theory suggests that particular 'accumulation systems' can be sustained through the medium term. A distinctive period of sustained accumulation is referred to as a 'regime of accumulation', which Boyer defines in these terms:

> The ensemble of regularities that ensure a general and relatively coherent progression of the accumulation process. The coherent whole absorbs or temporarily delays the distortions and disequilibria born out of the accumulation process itself.
>
> (Boyer, 1990:461)

Jessop explains the concept in these terms:

> An accumulation regime comprises a particular pattern of produc-
> tion and consumption considered in abstraction from the existence of
> national economies which can be reproduced over time despite its
> conflictual tendencies . . . relatively stable regimes of accumulation
> and national modes of growth involve a contingent, historically con-
> stituted, and societally reproduced correspondence between patterns
> of production and consumption.
>
> (Jessop, 1990:174)

Within any regime of accumulation a particular accumulation system is
necessarily supported by a mode of regulation through which individual
agents and groups collectively adjust their decisions and actions to a pattern
commensurate with the needs and constraints of the economy as a whole.
According to Boyer, the term 'modes of regulation' designates:

> Any set of procedures and individual and collective behaviours that
> serve to: reproduce fundamental social relations through the com-
> bination of historically determined institutional forms; support and
> steer the prevailing regime of accumulation; and ensure the com-
> patibility over time of a set of decentralised decisions, without the
> economic actors themselves having to internalize the adjustment
> principles governing the overall system.
>
> (Boyer, 1990:42).

Jessop offers the following definition:

> A mode of regulation refers to an institutional ensemble and com-
> plex of norms which can secure capitalist reproduction *pro tempore*
> despite the conflictual and antagonistic character of capitalist social
> relations.
>
> (Jessop, 1990:174)

In practice modes of social regulation necessarily encompass elements
which range from concrete institutional structures (such as legislation) to
intangible determinants of social action (such as values and norms of
behaviour). Peck and Tickell (1992:6) suggest that the analysis of modes
of social regulation might be usefully formalised in terms of five levels of
abstraction (see figure 3.1). Whilst it is crucial to recognise that elements of
this typology possess neither potential nor meaning in isolation, unpacking
the anatomy of modes of regulation in this way may still allow progress to be
made in understanding both their constitution and their function.

Whilst the more concrete forms of regulation (essentially those towards a the mode of social regulation [MSR] represents the concept in its most abstract form, as a generalised theoretical structure abstracted from the concrete conditions experienced in individual nation-states (for example, competitive regulation, monopoly regulation).

b within each MSR, a certain set of regulatory functions must be dispensed in order for the accumulation system to be stabilized and reproduced (for example, the regulation of business relations, the formation of consumption norms).

c the regulatory system is a more concrete and geographically specific manifestation of the abstract MSR, typically (although not necessarily) articulated at the level of the nation-state (for example, US Keynesianism, Pax Britannica).

d regulatory functions are dispensed through the operation of regulatory mechanisms, specific to each regulation system, which are historically and geographically distinctive responses to the regulatory requirements of the accumulation system (for example, the mobilization of labour power, the codification of financial regulation).

e regulatory forms represent those concrete institutional structures through which regulatory mechanisms are realized, although there need not be a straightforward one-to-one correspondence between mechanism and form (for example, local states, legislative systems).

Figure 3.1 Modes of social regulation

Source: Peck and Tickell, 1992

the bottom of Peck and Tickell's typology) are almost by definition easier to identify and study, it is possible to argue that they can only be meaningfully understood and evaluated in conjunction with more abstract forms of regulation. Such a contention stands on several points. First, concrete forms of regulation are only legitimised and, it follows, are only ever truly effective if they are underpinned by accordant and complimentary social values. And perhaps even more significantly, less tangible forms of regulation can often be seen as higher order determinants of social action. 'Concrete' regulation, in the sense of legislation and the like, attempts to moderate patterns of behaviour which are largely defined by these higher order modes of regulation; it follows that the most effective way to regulate development may be through strategies which attempt to influence the institutions, values and norms which are embedded in all societies.

Thus far these approaches have been divorced from sustainable development debates. However, if as we suggested in chapter 2, progress in the theory and practice of sustainable development depends on a fuller understanding of the causality of unsustainable events, and particularly on understanding the 'real' causes of these events, a regulationist perspective achieves considerable significance, not least because it has implications for understanding the production and legitimation of the unsustainable. This perspective not only allows us to conceptualise unsustainable events as outcomes involving structures and tendencies as well as purely contingent factors, it also allows us to begin to identify what mechanisms may be significant and, importantly, to consider what conditions allow these to be realised as unsustainable events. Current approaches to sustainability usually objectify unsustainability in concrete terms and address events directly. The approach we advocate here moves beyond this, but if it is to have any real utility, we still need to identify, specifically, what we understand to be the causes of the unsustainable. We need to identify significant causal mechanisms and, even more importantly, the particular conditions in which these are activated. To date, this sort of clarity has seldom been achieved.

The suggestion that sustainable development will be built around changed social values has been widely espoused. Consider, for example, the *Caring for the Earth* report's prescriptions 'to adopt the ethic for living sustainably, people must re-examine their values and alter their behaviour. Society must promote values that support the new ethic and discourage those that are incompatible with a sustainable way of life' (IUCN *et al.*, 1991:11); or as Redclift (1992:32) suggests, 'the tortuous road to greater global responsibility is likely to be built on the daily lives of human subjects, and recognition that these lives involve choices of global proportions'. However, as Murdoch (1992:7) points out, 'this shift in values will not take place simply at the level of the individual but will be the outcome of institutional practices . . . we should see this as a social process'. Certainly, the validity of any particular instance of regulation is necessarily dependent on not only its own relevance to the mode of social regulation as a whole but also upon the validity of that whole. The point that the validity of any mode of social regulation lies in its integrity is of considerable significance to consideration of how objective regulation might be articulated. For example, whilst it is tempting to suggest that the most appropriate locus for intervention lies in the 'higher order' moments of Peck and Tickell's codification, such a suggestion cannot be totally valid – these elements of regulation have no more substance when considered in isolation than do 'lower order' elements of the typology. Moreover, the validity of all elements of regulation is uncertain and insecure. Modes of social regulation emerge and achieve validity through conflict and struggle rather than through objective promotion. And within this, particular regulatory mechanisms, and indeed modes of social regulation as a whole, are subject to constant change. Thus it might be argued that current sustainability debates represent part of a process through which a new, and more sustainable, mode of social regulation will evolve – but this hardly seems to be the case. Sustainability concerns remain peripheral because the validity of past and emergent modes of social regulation is defined by quite distinct criteria.

Regulation theory and human agency

Historically, regulation theorists have been centrally concerned with the social reproduction, for instance, of the wage relation, necessary for the creation and maintenance of a viable accumulation regime. Viewed in this way, the development of the Fordist mode of accumulation involved not only a transition to a pattern of mass production and mass consumption, but also to the development of a set of social institutions necessary for this to occur and be maintained through time, for example, through the adoption of collective bargaining, a state adjusted minimum wage and a welfare state. According to Leborgne and Lipietz (1988:266), the emergence of these institutions was not the direct result of the capitalist dynamic *per se*, rather they reflected the ability of economic agents to internalise the logic of the regime of accumulation 'by anticipating the success of their initiatives'. However, for the most part, modes of regulation are not seen as being intentional and objectively constructed. Rather they result from a dynamic and uncertain process of struggle and conflict. As Aglietta explains, regulation theory

> simply accepts that the class struggle produces, transforms and renews the social norms which make economic relationships intelligible. These relationships have conditions of validity which are narrowly limited by the persistence of the norms which give rise to them. At our present level of knowledge of the problems of social transformations, we can accept here that if the class struggle produces norms and laws which form the object of a theory of social regulation, it is itself beyond any 'law'. It can neither be assigned a limit, nor be confined by a determinism whose legitimacy could only be metaphysical. In a situation of historical crisis, all that a theory of regulation can do is note the conditions that make certain directions of evolution impossible, and detect the meaning of the actual transformations under way. Thereafter, however, the future remains open. Historical development is totally different from biological evolution in as much as it is governed neither by chance nor by a hereditary determinism. History is initiatory. But it is only possible to construct a theory of what is already initiated – which puts a decisive limit on the social sciences.
>
> (Aglietta, 1979:67)

Accordingly, we must accept that regulation theory prescribes a very circumscribed significance for human agency. That said, however, it does not totally preclude the potential for objective and efficacious strategy. As Jessop (1990:77) points out, 'even at high levels of abstraction, the basic forms of the capital relation do not determine the course of capital accumulation. For the latter also depends on a variety of social practices, institutions, norms and so forth'. Capitalism may have inviolable laws but it has a plurality of logics.

Regulation theory therefore suggests a particular relationship between economic structures and imperatives and social action. By rejecting the notion that the conditions needed for the functioning and progression of capitalism are created in some miraculous way as structures reproduce themselves quasi-automatically without effective social agency, social action is ceded a certain consequence by regulation theorists. Boyer puts it thus:

> In stressing the structurally invariant features of the capitalist mode of production, one neglects to analyse the changes which mark it. One underestimates its contradictions, to the point where history appears almost immobile. In reality, it is a spiral, a process of innovation and of reproduction according to modalities that change from one period to the next.
>
> (Boyer, 1990:34)

Jessop comments on the situation in this way:

> In principle, its advocates refuse to study regulation in terms of a structuralist model of reproduction or a voluntarist model of intentional action. The reproduction of capitalist societies is neither a fateful necessity nor a wilful contingency. . . . The structure/ strategy dialectic does not separate struggle from structures but shows their complex forms of interaction. Structures are only prior to struggle in the sense that struggles always occur in specific conjunctures.
>
> (Jessop, 1990:194)

Clarke makes a similar point:

> The proponents of the model and Lipietz in particular, vehemently deny that it is either voluntarist or functionalist. There is neither a subjective will nor an inevitable logic underlying the emergence of a new regime of accumulation. Thus the phase of disintegration is a phase of class and political struggles which may be long drawn out, involving a lot of trial and error, before by luck rather than judgement, a stable regime of accumulation emerges.
>
> (Clarke, 1988:68)

According to Jessop, the related question of whether objects of regulation can pre-exist specific modes of regulation is a central issue. The various regulationist 'schools' which exist differ in their emphasis and hence in their conclusions on this point. Jessop (1990), however, suggests that the genesis of new modes of regulation is historically contingent rather than capitalistically preordained. For example, whilst he argues that 'capitalism cannot be understood without exploring the ramifications of the value-form', he is not suggesting a structurally defined, deterministic progression. Rather he proposes, albeit implicitly, a role for properly conceived human agency as he continues:

48

The substantive unity and expanded reproduction of the capital relation depend on successful co-ordination of different moments within
the limits of the value form. . . . In short, while the value form
defines the basic parameters of capitalism, neither its nature nor its
dynamic can be fully defined in value theoretical terms and further
determinations must be introduced. But once one begins to explore
how the value form acquires a measure of substantive unity, there are
many ways in which this can occur. Moreover, since capitalism is
underdetermined by the value form, each mode of regulation compatible with continued reproduction will impart its own distinctive
structure and dynamic to the circuit of capital. This implies
that there is no single unambiguous 'logic of capital' but, rather, a
number of such logics.

(Jessop, 1990:187)

In this sense, the capitalist dynamic is not strictly deterministic and a
'number of such logics' are conceivably valid. This permits the opportunity
for more sustainable modes of regulation and there is no reason to suppose
that different and more sustainable modes of development are not
tenable. But it is clear that these cannot be viable unless they embody the
'substantive unity' which Jessop describes.

While historically environmental concerns have existed largely outside the
capitalist development dynamic, the significance of the nature of the sustainability of economic development is now widely recognised (see Pearce 1995).
Indeed the idea of sustainable development is centrally concerned with the
fundamental and inextricable linkages between the economic, social and
environmental factors – the question is how we understand this whole. It is
clear enough, if we attempt to do this in ways which neglect the essential
unity involved, we relinquish the very essence of the idea. To date, this unity
has not always been respected and the theory and practice of sustainable
development has reached something of an impasse. The way beyond
this impasse is to explore the political economy of sustainable development
more deeply.

The origin of the unsustainable and the object of regulation: contradiction and crisis

Historically the environment has been as marginalised in regulationist thinking as it has been in other strands of social theory. That said, close parallels
exist between the concerns of regulation theory and issues which are central
to sustainability. Not the least of these is the fact that the contradictions
which ecological unsustainability pose to the continued viability of economic
regimes can be seen as being closely analogous to the other crisis inducing
tendencies with which regulation theory has been concerned to date. Thus

far, regulation theory has focused on crises which are endogenously derived, crises which are internal to the functioning of a capitalist economy: those which are the organic product of the system. It has prioritised labour and state relations over and above environmental ones. Environmental problems have been seen and addressed as little more than constraints on economic development – factors effectively external to the dynamic. However, although crises engendered by the constraints which the natural environment place on capitalist economies might appear to be exogenously derived, such a perception is superficial, if not erroneous. A broader conception of capitalist production would suggest that such exigencies may well be integral to the nature of capitalism. In reality, it may well be that a more rigorous and incisive analysis would suggest that the tendency to generate ecologically based crises is, in essence, very little different to any other propensity such as the widely studied tendencies to a falling rate of profit and to overaccumulation. As Smith puts it:

> In its uncontrolled drive for universality, capitalism creates new bar-
> riers to its own future. It creates scarcity of needed resources,
> impoverishes the quality of those resources not yet devoured, breeds
> new diseases, develops a nuclear technology that threatens the future
> of all humanity, pollutes the entire environment that we must all
> consume in order to reproduce, and in the daily work process it
> threatened the very existence of those who produce the vital social
> wealth.
>
> (Smith, 1984:59)

There are, however, ways in which ecological barriers to capitalist production might differ from the types of crisis with which most regulationist
analysis has traditionally been concerned. For example, however traumatic
and disruptive crises in the accumulation process of the latter kind may be,
ecological crises may prove to be much more fundamental and even more
traumatic. Some ecological crises are potentially so basic that it would not be
sufficient to merely postpone them as one might postpone the need to
devalue fixed capital. Moreover, in general terms the state has been somewhat
tardy in its treatment of these crises. Where it has acted, it has done so on the
basis of the severity of the outcome rather than by identifying and addressing the underlying cause. In effect it has seldom strayed from a neo-liberal
position, despite the opportunities to establish different types of solution.
Whilst the transition from one relatively stable period of accumulation to
another (for example, from fordism to post-fordism), may be disruptive, the
failure of the former does not in itself preclude the formation of the latter,
whereas this may be the case with some types of ecologically derived crises.
In this sense then, the types of crisis with which sustainable development is
concerned are sometimes quite radically different from the medium-term

avoidance of crisis which has characterised regulationist thinking. Although there may well be, as Moulaert and Swyngedouw (1989) point out, a spatial element within this, a transition from one regime of accumulation to another essentially involves a reconfiguration of the internal structures of the system rather than the redefinition of the limits of that system. At least, such a reconfiguration is sufficient for the transition to take place.

Exigency, expediency and expendability

Regulation theory is founded on the premise that capitalist socio-economic formations tend to be crisis prone and inherently unsustainable over time and space. Particular capitals and the patterns of social relations associated with these tend to become less and less viable through time. However, particular socio-economic formations can be, and are, more or less purposely sustained in the medium term despite the crisis prone nature of the capitalist mode of production. In order to sustain such formations, the contradictions and crises which threaten them are addressed through strategies which seek to maintain the viability of the status quo. These strategies are selectively legitimated and empowered by the 'mode of social regulation'.

Modes of social regulation encompass institutions, structures and values which act as channels between the real and the actual through which the dynamism of capitalist development is regulated and within which the nature of specific events is conditioned. Although our theoretical assertions have tended to obscure this, these implicate the environment directly and indirectly. They define rights, constraints, opportunities and powers which in turn influence the ways in which real causal mechanisms are expressed in practice. They serve to license and to some extent direct the nature of development. In capitalist societies, 'regulation' has been centrally concerned to maintain and control the value of capital and fixed assets, and it follows, with preserving and reproducing the existing power structures within society. This very process has predicated the unsustainable as strategies promoted to address internally generated contradictions have necessarily involved increasingly exploitative practices and the implicit redefinition of resources in ways which deny their true social value and perpetuity.

Figure 3.2 shows how the contradictory and crisis prone nature of social formations tends to be translated into a range of materially and morally unsustainable outcomes. As much of the twentieth century testifies, the inherent unsustainability of socio-economic formations can be postponed, but in practice only through measures which tend to involve increasingly profound forms of exploitation. A useful conceptual distinction arises here between what might be termed *relational sustainability* and *material sustainability*. The former is both the overriding object of regulation in capitalist societies and the condition which ensures the viability of a particular mode of social regulation. The latter encompasses the material and moral objectives of

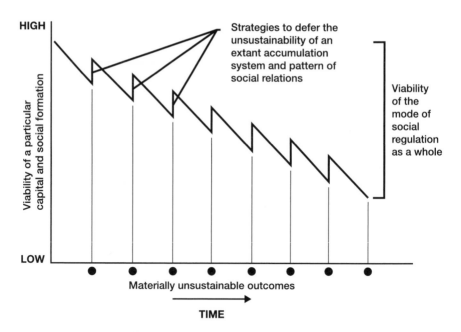

Figure 3.2 The mode of social regulation and unsustainable outcomes

sustainable development. What tends to occur in practice is that something which in itself is essentially inconsequential – *relational unsustainability* – is postponed, but only through processes which involve other more significant forms of *material unsustainability*. This process of translation is fundamentally conditioned by the nature of the mode of social regulation because it is this which selectively enables and empowers the mechanisms involved. This process of selection is biased. As they are currently constituted, modes of social regulation condition development in ways which make material unsustainability the norm.

As the viability of a particular socio-economic formation becomes threatened, strategies designed to preserve the value of capital and the viability of extant patterns of social relations are devised and promoted. Contradictions which emerge in a particular place at a particular time are deferred, for example through the provision of credit, or exported, for example through the exploitation of new resources and markets. Which strategies are actually 'successful' is determined by the mode of social regulation which selectively legitimates and empowers some strategies whilst invalidating others. However, the effectiveness of even 'successful' strategies can only ever be temporary as new and more profound contradictions will always tend to emerge. Thus it follows that subsequent strategies will necessarily involve ever more extreme forms of exploitation. The key point here is that however sustainability is defined, 'successful' strategies will inevitably tend to involve materially unsustainable forms of exploitation. Wherever the line is drawn, development will always cross it. While the mode of social regulation as a whole remains viable, new and more profound contradictions will always tend to emerge. And because these contradictions are more profound, effective responses will similarly need to become more and more exploitative. Measures which sustain particular capitalist formations in this way will always tend to produce outcomes that are unsustainable because they necessarily involve the progressively severe exploitation of both natural capital and of some groups or members of society. In effect, sooner or later, the line between sustainability and unsustainability will be crossed. And it will continue to be crossed so long as the mode of social regulation as a whole remains viable.

The overriding factor governing the validity of the mode of social regulation as a whole – *the* object regulation – is the effectiveness with which this whole can sustain the value of capital in the face of contradictory tendencies. A mode of social regulation exists in order to ensure the conditions needed for capital accumulation and because it is temporarily successful in achieving this. Given the centrality of this object of regulation, it becomes inevitable that modes of social regulation will tend to legitimate practices which produce materially and morally unsustainable outcomes. Notwithstanding expedients, such as the provision of credit and the application of new technologies, etc., regulation is always likely to reach a point where the only viable fixes involve the devaluation of natural resources or the lives of many members of society. It is usually a matter of where and when this occurs. Particular instances of regulation may, temporarily, postpone the expression of economic dysfunction and crisis (as for instance in the post-war Keynesian experiment), but the competence of such measures is always going to be limited – all such measures are at best temporary fixes. Hence in the twentieth century modes of social regulation tended to become increasingly based on the legitimate exploitation of environmental conditions – on a situation of increasing unsustainability.

Whilst particular elements of regulation or regulatory forms may avert particular crises, they tend to only redirect rather than counteract the tendencies which give rise to the events involved. An instance of regulation which is effective in that it serves to sustain a particular formation will almost inevitably transfer the problem to a different location or change the form of its expression. The more successful regulation is in perpetuating a particular accumulation system through time, the more extreme and more exploitative and potentially damaging the accumulation process is going to become. Without regulation the accumulation process cannot function, but inevitably there comes a point where continued accumulation can only be maintained through systems of exploitation which are by their nature *unsustainable*. The logic of this is that the longer any particular regime is maintained the more it will depend on modes of social regulation which legitimate materially unsustainable forms of exploitation. For example, the provision of credit at a variety of scales may well defer a crisis of over-accumulation. But, as with other elements of regulation, this type of strategy is merely an imperfect and temporary expedient rather than the basis of a truly sustainable system. Consider, for example, the ways in which high levels of international debt have tended to produce a whole range of unsustainable outcomes in the South. Past and existing forms of regulation have managed to operate on assumptions of unsustainability, if not in the same place then in others and at the expense of others. Capitalist accumulation and sustainable development have been and remain dialectically related. The overriding object of regulation exists in capitalist societies as the antithesis of the material and moral basis of sustainable development. However, thus far analysis of capitalist development has not sufficiently incorporated the progressive devaluation of nature through the reconstruction of regressive modes of social regulation, and debates about sustainable development have progressed as if these interconnections and dialectics are inconsequential for the concept and its reality. Indeed, despite the rise of a sustainable development discourse, if not a mode of development which embodies it, the tendency has been to assume that it can be bolted on to existing environmental and social relationships and the real(ist) conditions of sustainability have been obscured.

The regulation of sustainable development

These propositions suggest that a move towards sustainable development requires a reconstitution of the regulatory mode in ways which change this central object of regulation. As they are currently constituted modes of social regulation prioritise the value of capital and existing class structures while incidentally marginalising the material basis of sustainability. By defining the object of regulation in this way society legitimates and empowers a set of causal mechanisms which sustain wealth and privilege at the expense of a whole range of materially and morally unsustainable outcomes. This need not be the case, but purposive change depends on a more developed understanding of the mechanisms and conditions which create and reproduce this situation than currently exists.

There is no single uniquely deterministic logic of capital which defines the object of regulation in this way. We should remember that modes of social regulation are just that – social. They are socially produced and reproduced. They can be changed. That said, it is important to recognise that they are all-embracing and almost infinitely complex, and, crucially, that they come about and achieve validity through a process of experimentation, conflict and struggle rather than through any form of objective promotion. Society cannot simply construct new modes of social regulation as valid wholes. But what is being suggested here is not that modes of social regulation can be objectively constructed *per se*. Rather that the core values and institutions which legitimate and empower the mechanisms which underpin unsustainable outcomes can be changed. A key problem here is that any strategy to redefine the object of regulation in this way is necessarily radical in that it challenges the existing social order. Accordingly, the fact that regulation is normally realised through existing power structures appears to represent a major barrier to the promotion of any such agenda. This aside, it seems clear that sustainability can only be built around institutional and value shifts in society, and, moreover, that it is not simply the values which society places on environmental resources and human lives which are important. Equally, if not more, significant are the values and institutions which prioritise the value of capital and the maintenance of existing patterns of social relations. These core values can be changed. However, if new modes of social regulation are to emerge, if viable accumulation systems are to incorporate natural resources in new ways, the stimulus for this needs to emerge from progressive breakdowns of struggles dynamically experienced within the mode of social regulation – which becomes an important analytical focus. And within this analysis, it is crucial that unsustainable practices are understood and addressed as outcomes rather than as events *per se*.

Strategy must be formulated within a conception of sustainability which recognises the transformational nature of capitalism. Historically, and today, the object of regulation and the viability of any regulatory mechanism are dependent on their relevance to the existing accumulation process. From this perspective, if key elements of sustainability are to transcend particular regimes of accumulation, there can be little utility in merely sustaining the value of capital or an established pattern of social relations. Such strategies are inevitably tactical and ultimately untenable exercises in extemporisation. But more than this, they ultimately promote the destructive over-exploitation of both nature and labour. The achievement of sustainable development requires that the object and nature of regulation are extended to incorporate environmental and moral criteria. Sustainability needs to be articulated in the reflexive progression of capitalism and the social systems which sustain and renew the dynamism of capitalist accumulation.

A central tenet of the regulationist perspective is that the social, institutional and economic regularities which constitute any particular regime of accumulation are 'unsustainable' in the sense that they are inevitably temporary. The collapse of any particular regime of accumulation will necessarily tend to occur because of internally generated contradictions which progressively engender incongruence and disequilibrium. In itself this is not a problem. A new set of regularities, a new quasi-stable form can emerge phoenix-like from the ashes of the old. Indeed the old must go before the new can emerge. But here, surely, lies at least part of the explanation of why the unsustainable occurs. The emergence of a new regime of accumulation necessarily involves the devalorisation of the old. The problem here is that whilst the devalorisation of the pattern of relationships existing in the old would be sufficient to allow the new to emerge, these are only devalued when regulation has failed. In practice, what tends to occur is that disequilibria which emerge in established socio-economic formations are ameliorated through increasingly severe processes of exploitation involving expropriation and devalorisation of resources both within and outside the system itself. In this way the inviability of structures internal to the capitalist system itself, for example particular class structures, are translated into true unsustain-ability. The internal contradictions of the capitalist system become, through regulation, externalised in ways which produce material and moral forms of unsustainability.

The achievement of sustainability requires that the internally derived dis-equilibrating tendencies of capitalist systems, including a tendency to materially unsustainable outcomes, are understood and addressed in ways which are deep and powerful enough that their expression can be moderated. If this is to be achieved it will necessarily involve the construction of new and different modes of social regulation. If these are to be promoted, a crucial significance attaches to understanding how the essentially internal is trans-lated into the external, and within this, how this process is validated and empowered through particular regulatory processes and modes of social regu-lation. If this can be achieved, it may then be possible to devise and put in place new modes of regulation which do not, by virtue of what they are, inevitably promote the unsustainable. As they are currently constituted, modes of social regulation condition development to the unsustainable. This conditioning is what needs to be moderated if sustainable development is to be achieved. It may not be possible to 'manage' development, but that is not to say that it cannot be regulated in new and different ways.

The regulation of sustainable development: a research agenda

As the impossibility of managing sustainable development has become increasingly apparent, a degree of consensus has emerged concerning the need to move away from a focus on eventual outcomes to consideration of why and how these are produced and reproduced by underlying social pro-cesses and conditions (Dickens, 1992; Benton, 1994; Jacobs, 1994). From this perspective, the way ahead does not lie in finding more 'objective' methods of defining what is or is not sustainable in any particular case. Rather, the need is to better understand why overly exploitative and degrad-ing practices come about and how they are able to achieve a large degree of social and political legitimacy. This requires that a conceptual framework and methodology for articulating sustainable development is developed through a closer engagement with social theory. Within this, it is particu-larly important that research explores the ways and potentialities of present systems of economic and social regulation assessing how institutional and value change at the social and economic level might be effectively promoted. Insights from critical realism and a regulationist perspective on the nature and dynamics of capitalist economies may thus play a major role in defining a more productive approach to progressing both the theory and practice of sustainable development.

Realism provides an ontological and epistemological basis for understanding the causality of unsustainable practices and events. Regulation theory is potentially useful in that it can begin to clarify our ideas as to what can and cannot be sustained. Although a regulationist perspective posits a highly circumscribed potential for human agency (Jessop, 1990; Jessop, 1995), it can also suggest how we might seek to condition a more sustainable future through the construction of new and different modes of social regulation. Understanding the role of modes of social regulation and the conditions which ensure their validity may thus be important in understanding how sustainable development might be promoted. If such 'reconditioning' is to be attempted, the first step is to understand sustainable development as a condition rather than a criterion through which development can be evaluated and managed. Certainly the unsustainability which pervades present day modes of development is constituted in specific practices and events, but these events are predicated by structural elements of causation and the conditions through which the tendencies which these give rise to are mediated. A key point here is that the outcomes produced are not 'determined' by these structures and conditions; rather these factors mean that particular types of outcome are always likely to be realised.

Historically, 'regulation' has been centrally concerned to maintain the value of capital and to preserve extant patterns of social relations in the face of contradictory tendencies. Thus 'regulation' has tended to legitimate and actualise increasingly profound forms of exploitation, and, sooner or later, this exploitation tends to cross the line into unsustainability. The achievement of sustainable development requires that the criteria which define the validity of modes of social regulation are themselves expanded. If this expansion is to be promoted, an understanding of what must be regulated and how and at what level this might best be achieved becomes crucial. If viable accumulation systems are to incorporate natural resources in new ways, the stimulus for this needs to emerge from progressive breakdowns of struggles dynamically experienced within the mode of social regulation.

The subsequent chapters of this volume revolve around the development and application of the research agenda and methodology outlined above. Accordingly, the project was concerned to explore the ways in which the subjectively formulated strategies of actors within two industries interacted with the structural dynamics of capitalist accumulation in a process of experimentation and struggle; and to test the contention that the outcomes of these struggles are biased in ways which condition development to the unsustainable. Whilst established realist methodology places considerable emphasis on the 'unpacking' of events in order that causal mechanisms can be identified and evaluated, this research was equally concerned to understand how particular modes of social regulation selectively legitimate and 'activate' these mechanisms. The challenge for the project was to situate and interpret 'concrete' instances of unsustainability within a model embodying the multi-layered mode of explanation defined by a realist ontology.

The research undertaken focused on two case studies of sugar cane production: one in Barbados and the other in the Australian State of Queensland. The sugar sector was selected for two principal reasons. First, it embraces many of the environmental, economic, social and moral concerns which are widely held to be significant to sustainable development. Second, the nature of the global sugar economy, and the relatively long period through which sugar has been produced in the case study locations, provided the potential for a productive analysis of the relationship between the dynamics of capitalist accumulation and sustainability concerns.

The choice of Barbados and Australia as particular case studies within the sugar sector was based on a number of factors. Given the nature of the realist research process outlined earlier in this chapter, it was anticipated that a comparative study of this type would be useful in that it would facilitate the identification of significant elements of causality within situations made complex and unclear by contingent factors. There are broad areas of commonality between these two case studies, not the least of which being that both locations produce an essentially identical commodity – sugar. There are, however, also significant differences between the sugar industries in the two study areas. Australia is a highly developed country, whereas Barbados, whilst it hardly has the problems of some developing countries, remains part of the 'South'. The Barbadian sugar industry, which has existed for over three hundred years, is now on the verge of total collapse, whereas the Australian sugar industry is often held to be a paragon of efficiency, innovation and, implicitly at least, of 'sustainability'. Another key difference lies in the fact that Barbadian sugar production has remained plantation based, whereas in Queensland production has been based on family farms for almost one hundred years. Over and above these differences, the ways in which the sugar industries in these two locations are regulated are also very different. This is true not only of the more concrete forms of regulation which exist in these two locations, but also with respect to the less tangible elements of their respective modes of social regulation.

Realist methodology requires that a model informed by both actual events and theoretical constructs is progressively constructed, substantiated and improved. The question of where the researcher 'breaks into' this model in order to begin the process of conceptualisation and reconceptualisation is not particularly significant from a methodological perspective. In practice, the research described in this volume initially evolved around key theoretical categories derived from regulation theory and a preliminary analysis of the nature and history of the global sugar economy. The history of the two case study industries was also explored at this stage using data from a range of secondary and primary sources including official publications, academic texts and the records of industry bodies. In so much as established realist methodology suggests that the model being tested and refined should be progressed in ways which inform the original conceptual constructs, this methodology was not discordant with established definitions of what is appropriate in realist research. Somewhat more problematic, however, were the criteria used to define 'unsustainable' events which formed the other domain of the model being tested.

The central tenet of the argument being developed here is that attempts to objectively define what precisely constitutes sustainable or unsustainable development in any particular place and time are ill-conceived. However, in so much as the model being tested and refined required 'concrete' instances of unsustainability, two basic definitions were used. First, much literature addressing the two case studies (for example, Caribbean Conservation Association *et al.*, 1994; and Department of Primary Industries, 1994) defined sustainable development in their own terms. Similarly, many of the individuals interviewed during the research also had their own interpretations of the concept. Second, it was assumed that the externalisation of contradiction and dysfunction emerging within any particular production system constituted a form of unsustainability. Thus for example, water extraction from an aquifer which exceeded the rate of replenishment, or farming practices which involve highly polluted run-off were held to be 'unsustainable' for the purposes of this research. These criteria may be inexact and contestable, but they were appropriate and sufficient. The fundamental point is that, in themselves, the events which were actually used as examples of unsustainability in the research and subsequent analysis have little significance to our central argument. They may or may not constitute examples of unsustainability, but this is in no way consequential. To have sought any 'objective' definition would have been fruitless, our objective has been to move beyond the nihilistic logic of such 'objectivity'.

The main empirical component of this research revolved around a series of interviews conducted in Barbados and Australia. A particular problem with research which attempts to construct a model of causality through interviews with individuals is that their interpretations of causality are individually and collectively subjective. Certainly, regulation theory posits a very circumscribed potential to human agency; and it has been argued that it can provide little more than a context within which development can be interpreted:

> At best we have more or less plausible regulationist conceptualizations of these shifts. Yet, however detailed the analysis of the

strategic context might be, it cannot itself generate an adequate explanation for strategic action. This would require in addition at least some account of the strategic capacities of actors (individual or collective) to respond to economic problems, the strategies they pursue and the relationship between these capacities and the strategies and those of other relevant actors in that context.

(Jessop 1995:321)

From a regulationist perspective, actors may anticipate the success of particular initiatives (Leborgne and Lipietz, 1988:266), but their consequence within the processes of struggle which constitute the actuality of development cannot be ensured. Actors within and around the Barbadian and Australian sugar sectors are clearly significant in that their actions serve to reproduce and sometimes transform existing structures. However, their strategies are necessarily formulated within bounded rationalities and influenced by the conditions in which they are articulated, and in practice the reproduction and transformations which have occurred have often been less than intentional.

Structures are seen as durable, sometimes capable of causing social change or social conditions, and also capable of locking their occupants into role positions. They are often difficult to displace and transform, yet are continually reproduced by the actions of people, who in turn are often not reproducing structures in any way intentionally.

(Cloke *et al.*, 1991:150)

As Bhaskar suggests:

People do not marry to reproduce the nuclear family or work to reproduce the capitalist economy. Yet it is nevertheless the unintended consequence (and inexorable result) of, as it is also a necessary condition for, their activity.

(Bhaskar, 1979:44)

In terms of this research, the unquestioning faith in modernisation which pervades the Australian sugar farming sector is often just that – unquestioning, but it is still potentially causally significant. This does not mean that actors should be regarded as 'cultural dupes, programmed to perform roles and reproduce structures' (Cloke *et al.*, 1991:150). Rather, the rationality and significance of strategy needs to be interpreted *in context*. From a realist perspective, it is also important to recognise that individuals have causal powers and liabilities in much the same way as inanimate objects and that these are activated by contingent factors (Sayer, 1984). Sayer's interpretation on the implications of this are summarised by Barnes:

Sayer makes three further claims: (1) that unlike inanimate objects, individuals have the ability to learn and thus the power to change their causal powers and liabilities over time; (2) that while such change is possible, the intersubjective beliefs that constitute subjectivity are relatively stable and are reproduced through a recursive relationship between the individual agent and the broader social structure; and (3) that in order to understand the subjectivity of agents and hence the causes of their action, it is necessary for the researcher to engage in some form of interpretative understanding, or *verstehen*.

(Barnes, 1996:20)

Questions of rationality and subjectivity are important to thinking on sustainability. Thus far, most approaches to sustainable development have assumed that sustainability goals can be rationally promoted through specific initiatives. This has proved to be an overly ambitious assumption. This book is an attempt to consider how sustainable development might be promoted despite the limited rationality and scope of human agency. Thus although as Jessop (1995) suggests it may well be the case that 'detailed analysis of the strategic context' cannot itself generate an 'adequate explanation for strategic action', a realist epistemology allows us to address this problem by considering more fully the significance of the context itself.

Derived research methods

The critical realist mode of explanation places considerable significance on the context in which development occurs, and a key objective of this project was to consider the ways in which particular conditions effect the outcomes produced. The research was concerned to elucidate struggles experienced within the political economies of the two case studies, and to consider, in particular, how the sometimes intentional but often less than deliberate reproduction of key structures has frequently served to produce unsustainable outcomes. More precisely, the objective of the research was to test and refine the model at the heart of the thesis (figure 3.2) through its application to the case study industries. This was achieved, first, through an historical analysis of the two case studies and, second, by relating the model to present day situations.

The historical analysis revolved around a number of key texts including Deerr (1949), Blume (1985), Mintz (1985), Coote (1987) and Abbott (1990). This was complemented by an analysis of more specific secondary data sources covering the development of the sugar industries in both Barbados (for example, Watts, 1987 and Beckles, 1990) and Queensland (for example, Saunders, 1982; Manning, 1983; Kerr, 1988 and Graves, 1993). A range of primary sources were also used to provide background information on the two case study industries. The central significance which sugar held in the Barbadian economy until recently has meant that some aspects of the industry's development such as acreages, yields and prices, have been quite well documented, for example in the records of various sugar mills and in a variety of official reports which have been commissioned into the industry (McGregor *et al.*, 1979; Booker Tate, 1993). Similarly, the highly regulated nature of Australian sugar production has meant that equivalent data were relatively accessible (ABARE, 1985; SCIST, 1989; ABARE, 1991). In both these cases, however, the amount of published qualitative data is less comprehensive.

The empirical component of the project was designed and conducted in accordance with established definitions of what constitutes an appropriate realist methodology. The concern was not to survey taxonomically defined or representative groups or individuals, but rather to focus on causally significant groups (Sayer, 1984). Interviews were conducted with a range of individuals in both Barbados and Australia, with these individuals being selected from groups who appeared to relate to each other either structurally or causally rather than because of any aggregate formal relations amongst taxonomic classes (Whatmore, 1995:33). In both case studies, an attempt was made to interview individuals both within and outside the sugar sector itself and to explore the interviewees' own interpretations of events, relationships and conditions. The need to extend the interview process beyond actors directly involved in the two sugar industries concerned was seen as important because of the potential significance of the broader context in which the two industries operate. For example, the general antipathy to sugar production which exists in the Caribbean as a legacy of slavery (Beckles, 1990) may well have a significant influence on present day events.

In accordance with Sayer's (1984:223) suggestion that respondents may be selected 'one by one as an understanding of the membership of a causal group is built up', the selection of interviewees beyond the relatively small number initially targeted was effectively determined by the nature of the early results of the research, which served to elucidate apparently significant causal relationships within and between groups. Although such a research design may appear to resemble a directionless 'fishing expedition', it is accordant with realist concerns for 'explanatory penetration':

> it is possible, though not mandatory! – for intensive research to be exploratory in a strong sense. Instead of specifying the entire research design and who or what we are going to study in advance, we can, to a certain extent, establish this as we go along, as learning about one object or from one contact leads to others with whom they are linked, so that we build up a picture of the structures and causal groups of which they are a part.
>
> (Sayer, 1984:244)

In practice, 44 taped interviews were conducted in Barbados and 47 in Queensland during the first six months of 1994. A number of discussion groups involving farmers and representatives of the Canegrowers Association were also held in Australia. In each of these locations, a wide range of actors were interviewed. These included: planters, farmers, mill workers, managers of sugar multinationals, estate workers, and individuals involved in the wider regulation of the industry – politicians, agriculture ministry officials, extension workers. Little difficulty was experienced in gaining access to appropriate interviewees in either Barbados or Queensland. Only two of the potential respondents approached in Barbados declined to be interviewed. Similarly, although Australia is a large country, the sugar producing areas are relatively small and those involved in the industry tend to be open and usually quite happy to discuss their involvement with and perceptions of the industry.

In Barbados, a preliminary analysis of the literature allowed apparently significant 'causal groups' such as plantation owners, small farmers, the government, the financial sector, and labour interests as articulated through the Barbados Workers Union, to be identified before the empirical phase of the research began. In practice, members of each of these groups were interviewed early in the research process. However, what emerged was a situation in which the initially identified groups were not particularly meaningful. The nature of what were in fact causally significant groups emerged as the research proceeded, for example, as the lack of distinction between apparently different groups such as the planters, government and various regulatory bodies became more evident. In practice, this lack of distinction is a significant element of the conditions in which Barbadian development takes place. Thus while subsequent interviews involved similar individuals, a new and more meaningful understanding of how causal groups were constituted evolved throughout the research process. In Australia, it was similarly possible to target initial interviews on members of apparently significant causal groups which included farmers, the regulatory authorities, the financial sector, industry bodies such as the Canegrowers Association and the milling companies. Again it was possible to identify subsequent interviewees in the light of initial interviews. In this case, the groups were more clearly defined and meaningful, but the importance of factors such as the ethnically defined communities which exist in some parts of the Australian sugar farming sector only emerged during the research process.

As Sayer (1984:54) suggests, what is important to realist research is 'learning from the respondents what the different significances of circumstances mean for them'. And in accordance with this definition of realist methodology, the interviews, although tape recorded, were conducted in a very loosely structured manner and questions were kept as 'open' as possible. Although the interviews were largely unstructured, the aim of understanding causality in realist terms was used to steer the conversations. The interviews were not conducted in order to specify what is or is not sustainable (although in Australia, in particular, the vast majority of respondents were quite familiar with the term). Rather, they were conducted with the quite different objective of testing and refining the model outlined in figure 3.2.

As the literature suggests is appropriate (see, for example, Pratt, 1995), the research progressed iteratively as the initial model book was tested and refined. In both case studies, apparently significant causal mechanisms were identified and substantiated from a realist perspective. In practice, the identity and primacy of particular mechanisms tended to emerge during the empirical component of the project and subsequent interviews were focused to explore these. Similarly, the nature and significance of the contexts in which the Barbadian and Australian sugar industries are developing also became clearer during the course of the empirical research and again an effort was made to focus discussion around these factors.

The final phase of the research involved a reconsideration of interview transcripts and primary data collected during the fieldwork and a reinterpretation of literature reviewed earlier in the research process. In practice, it was possible to identify and, to some extent at least, substantiate a number of apparently significant causal mechanisms and to begin to describe the significance of certain conditions. This, in turn, allowed further testing and refinement of the theoretical categories and model defined early in the research process.

4

INTERNATIONAL FOOD SYSTEMS
Creating spaces for foods and nature

Introduction

The production, supply, manufacture, retailing and final consumption of food represents one of the most significant spheres for the regulation of capitalist economies both in the North and the South. It also holds profound implications for the environment. This chapter, serving as a preface to case studies concerning sugar production in two regions, examines the particular contribution studies of food are making to the understanding of regulation and environmental sustainability. In particular, following on from chapter 3 it will ask how specific modes of social regulation have developed, and how these then begin to engender types of development which are unsustainable. The struggle to incorporate food into the wider study of modern capitalist development has been a major feature of recent rural and agrarian sociology. In doing so much of the work has attempted to apply aspects of regulation theory, and to a lesser extent, realist ontology. Part of the work of this study has been to trace the resulting trends.

In tracing these trends, however, it has become apparent that there has been a failure to incorporate either the dynamic changes in social formations which are responsible for legitimating the growing contradictions inherent in food systems, or the relative significance of the natural properties of foods. Current meta-theoretical development concerning food and agriculture is thus at something of an impasse, for however much it attempts to connect with these socially constructed phenomena, its conceptual apparatus tends to restrict their significance to external and fragmentary factors. It is proposed, as our earlier propositions suggest, that it is necessary, in order to overcome these difficulties, that scholars reconsider the flexible time and space configurations of agricultural and food development, taking on board material sustainability concerns.

Systems and regimes : the adoption and development of regulation theory

The study of globalisation has become a major area of concern in the social sciences over the past decade. Not least, in the rural and agricultural sphere, it has been recognised that the growth of new global linkages in food trade, in their influence upon production methods, and in the transfer of specific knowledges about food, they have all been influenced by global processes such as the rise of transnational capital and the use of sophisticated transport and communication systems (see Bager, 1997). Like most profound intellectual developments such a movement has developed several significant strands.

Of particular importance, for instance, is the division between those scholars who tend to emphasise the most recent changes, focusing upon the contemporary global food landscape, compared with those who continue to stress the need to take a longer historical food systems perspective, putting a stronger emphasis upon the evolution of regimes of accumulation which have developed in line with, or at least as a significant element of, capitalist development more generally. The former strand, in its more critical of modes, has tended to jettison many of the regulationist assumptions of the latter (see Goodman and Watts, 1994; Marsden and Arce, 1995; Whatmore, 1995), preferring to centralise concepts of social action and contingency in the analysis of food and agriculture. This is leading to some exciting developments in integrating actor strategies with food chains and networks, and in examining the ways in which the relative engagement with agriculture and food incorporates space and locality. It emphasises the need to contextualise social action in local and national spaces and tends to downplay the significance of capitalist accumulation and commodity relations.

It is the latter strand, and in particular, how it may potentially interface with the former which is of interest here. While the former group of scholars have tended to disregard or side-step the regulationist and realist arguments implicit in the more systems and regime oriented work, the latter, it would seem, have also tended to forget much of the salience of the realist arguments outlined in chapter 2. As a result they have fallen (somewhat easy prey) to the social constructionist point that their ontology is far too structurally determined and machine-like on the one hand and globally reified on the other. Alternatively, as they have defended:

the concept of food regime is an historical concept, which is why it addresses geopolitical as opposed to geographical concerns. As a historical concept, it is also comparative – not geographical, but historically comparative. That is, it specifies the political history of capitalism, understood from the perspective of food. Thus the 'food regime' distinguishes two periods of recent capitalist history (late-19th century, mid-20th century), each of which is framed by contradictory principles of organisation, territorial/geo-political and capitalist-marking hegemonic transition.

(McMichael, 1996a:48)

As we shall see, this also tends to downplay a genuinely realist stance and fails to appreciate the role of modes of social regulation and formation in managing and modifying food regimes over space and time.

The thesis developed throughout this book takes neither abject stance. While it accepts the richness of both perspectives it posits a third way. One which reinstates a realist perspective in order to incorporate the condition of sustainability. It wishes to examine how a more nuanced realist and regulationist analysis can harness the most optimum explanation of change and, particularly, how that explanation of change can incorporate sustainable development.

In doing this, however, it is necessary to rehearse some of the basic features of food regime analysis and then to suggest ways in which these can be developed so as to accommodate and enhance sustainability debates. As we shall see, neither strand of the globalisation debates in rural social sciences have satisfactorily conceptually incorporated consideration of sustainability.

Food systems and regimes: without nature

Aspects of regulationist analysis, if not realist ontology, have formed a basis for agrarian sociology debates over the past 15 years. Their development arose out of the need to understand the recent historical rise of agro-chemical and mechanical technologies together with the advancement of genetic engineering in the production of plants and animals in the post-war period. This was a period, in McMichael's terms, of the developmental project. Technological advances which had constituted a significant means for changing the spatial advantages and disadvantages of agricultural production, and the differential role of state intervention at the national level, were seen as breaking down, giving way to wider financial, corporate and manufacturing global capitals. How these global corporations worked, how they were organised, and how the declining hegemony of nation-states were to cope with these trends became the touchstone for a reinvigorated agrarian sociology debate concerning the globalisation of agriculture and food, and more specifically, the systemic analysis of food production and consumption (see Goodman and Redclift, 1991; Bonanno *et al.*, 1994; McMichael, 1994, 1996b). Moreover, this also attempted to bridge the gaps between the North and the South, demonstrating how southern regions were tied into the global accumulation dynamic.

As several writers have now developed (see Marsden *et al.* 1993; Lowe *et al.* 1994) one particular feature of this work was the historically comparative analysis of food regimes at the global scale of analysis. Following generalised regulationist principles (if not those which Jessop and others have more recently espoused concerning actor strategies and institutional practices), the identification of broad periodisations has occurred, with the delineation of first, second and (possibly), third food regimes (Friedmann, 1993). Each of these has their own regimes of accumulation and social regulation, with the latter being associated with particular modes of production and consumption.

The first food regime (1870–1914) was one built upon an imperial (British) priority which prioritised domestic stimulation of manufactured goods 'in exchange' under so-called 'freemarket' conditions for the importation of an ever widening range of food raw materials (coffee, tea, sugar, wool, wine, lamb and mutton) from the colonies, under a system of imperial preferences. This was largely replaced in the 1920s and 1930s, after the continued growth of petty commodity production and agribusiness in 'settler countries', the development of state assisted technologies, and the development of the Fordist compromise which balanced the functional employment needs of urbanised industrial production with the requirements to produce relatively cheap foods for its industrial workforce.

This ushers in the intensive food regime whereby the necessary aggregate increases in food availability and agricultural production could only be achieved through the increasing intensity of production and manufacture of the foods themselves. Here the accumulation of capital occurs not so much through the use of extra land and labour resources, as through the continued adoption of new technologies which raise output per worker and hectare.

The analytical value of these periodisations is enhanced by the ease with which they relate to broader macro-economic and political structures, placing the food economy in the context of overall capitalist development (Le Heron, 1993). For instance, the second food regime is interactively related to the rise of New Deal politics ('a car in every garage and a chicken in every pot'), and the broader post-war advanced world consensus – for instance, as typified by the activities of the FAO and the Marshall Plan as well as most national governments in attempting to use the state to ensure the exponential rise in food output in the post-war period. These infrastructures represented the basis of a mode of social regulation which upheld what has become the super-intensive model of agricultural production and food supply. It has, as many of the commentators point out, developed globalised proportions and led to a reconfiguring of agrarian uneven development. The technological treadmill has been upheld by a progressive and modernising ideology. It has been particularly erosive in its treatment of peasant agricultures in the South, giving precedence to the manufacture of Northern countries' foodstuffs and the standardisation of food goods. For instance, advanced countries like the US became net exporters of primary and processed foods (e.g. soya beans) by the early 1980s. Between 1960 and 1980 its exports grew by 189 per cent representing half of all global production.

The highly regulated and state-supported mass markets created by this regime and the standardised products which pass through them, were now dominated by the North and were more favourable to their own agricultural producers and powerful transnational corporations. This set the capitalist rules of the game for Southern agricultures, placing many of them in technologically inferior positions and forcing more intensive efforts to export agricultural commodities at the expense of their domestic and self-sufficient needs.

Such periodisations and their theoretical linkages to aspects of wider society (such as the household division of labour, food politics, the industrial wage relation and the maintenance of particular nation-states as powerful actors over others), have tended to hide some of the unanswered questions concerning broader aspects of regulation theory and the use of realism as an ontology. We want to now deal with these issues in relation to the real ability of food regimes to deal with temporal and spatial change.

Food systems and regimes . . . with nature

Of course, we have to recognise that, as Lowe *et al.* (1994) argue:

> Although specific regimes are inductively derived and refer to particular historical periods, the general notion of an international food regime is a theoretical construct rather than an empirical category. It is useful as a heuristic framework, in ordering broad geographical and historical experiences and in directing research effort to critical periods and agents of transformation.
>
> (p. 9)

While this may be the case it is also important to identify some of those aspects of transformation it fails to deliver in explanatory terms. For instance, there is much less of a consensus about the character of a third food regime which is based around, for instance, the decline of plantation agricultures and the uneven rise of the new export agricultures in the South, following the partial deregulation of the post-war state infrastructure (following successive GATT rounds) and the rise of retail corporations as major players in global deregulation of food markets (see Marsden, 1997b). In addition, changes in mass and more specifically niche consumption of food in the North have at least cast something of a cloud over the notion of a 'Fordist' diet and the functionality of 'white goods' in the kitchen as harbingers of the further industrialisation of foods. Moreover, because of the lack of recourse to transformative change at the local and regional levels, it is not clear how one should interpret the obvious co-evolutionary development (what

O'Connor (1988) and Redclift (1990) term 'combined development'), where both aspects of the super-intensive and productivist second food regime and the more recent deregulated and more spatially specified 'third' regime coexist. The latter, built upon more diverse and consumer-led markets, more consumer niches, reconstituted class/consumption arrangements, less emphasis upon food manufacturing and more upon corporate retailing, and highlighting local and regional uneven development and dependent and dominant spaces (see Marsden, 1997a), is by no means temporally separated from the more industrialised model. Rather, they seem to be coevolving.

The reluctance to embrace the complex nature of combined and uneven development in much of the food systems literature has been justified in terms of the alternative need to historically account for the particular global geopolitical relations as they evolve in different broad periods. Empirical analysis has in these terms been regarded as the responsibility of at least two other strands of the globalisation of agriculture and food academic project. These have been concerned with either, what McMichael (1996b) rather disparagingly calls 'localistic studies' – a growing caucus of literature on the theoretically informed study of local rural spaces (see Murdoch and Marsden, 1995), or with the equally Anglo-American literature concerning the empirical analysis of commodity chains (see Friedland, 1994). Interestingly, both of these separate literatures, with the former belonging much more to the European rural sociological tradition and the latter to the American critical zest for tracing the complexities of corporate food manufacturing, have in their separate ways pointed to the areas of conceptual weakness in the more over-arching food systems and regimes literature. The mechanisms and the contingencies of corporate power begin to be exposed in the more recent post-GATT landscapes of the 1980s and 1990s. And the integrative nature of rural development processes in the local studies serve only to emphasise both the variability in the degree and direction to which the food economy affects different rural areas.

These fissures and different directions in the debates over the new or 'third' food regime are tending to hide the significance of many of the principles of realist and regulationist analysis which in many cases were responsible for guiding such bodies of work (see, for instance, Lawrence and Vanclay, 1994; and Burch et al. (1996) on the environmental implications for Australian agriculture).

As McMichael (1996b) concludes, rather than progressing a dichotomous path between social specificities and constructionism on the one hand or the structuralist potentiality for 'violence of abstraction' on the other (see Sayer, 1984), we need to reintegrate some of the key aspects of realist ontology into a more integrated approach to food and rural development.

> Unless we specify the historical relations in our concepts they remain abstract. 'Levels', or units of analysis cannot be taken as given. Social

units are self evident in neither space nor time ... they form relationally. In this sense the opposition of local and global analysis is a false opposition, as each template is a condition of the other. On their own, conceived in non-relational terms, global and local "units" can only exist in refined levels of analysis.

(p. 50)

A further evolving gap in the food systems literature, and one partly responsible for limiting its appeal at the current time, concerns the difficulties it has experienced in embracing nature, either as a political and economic form of 'capital' or as a stock of social resources upon which social groups or society as a whole confers different types of value. In much of the literature it is widely regarded that environmental concerns and 'questions of nature' need to be incorporated into agrarian political economy (see Marsden *et al.*, 1996). This is either seen as necessary in order to identify the more fuzzy architecture of the 'third' food regime, or used to further attack the systems theorists for failing to recognise the real natural and potentially distinctive character of food–society relations *vis-à-vis* other industrial and economic processes. Moreover, several writers have documented the environmental negatives of the second food regime in particular (see Lowe *et al.*, 1990). This contributes to much of the burgeoning political ecology literature which has been infused with the growing material unsustainability of agricultural practices in the South (see Blaikie, 1985; Blaikie and Brookfield, 1987). Despite these forays into the challenge of analytical incorporation of the natural, much of the agrarian political economy of food and the food system literatures has been content to render food commodities like any other, as products and inputs into a complex and increasingly global food economy.

Nevertheless, the recent emergence of the sustainability agenda tends to challenge these assumptions, and it calls for more than simply a 'surface' recognition of the ways in which nature is implicated into other aspects of social and economic development. Moreover, however unsustainable the evolving capitalist food regimes would seem to be, unless scholars begin to incorporate sustainability concepts they are unlikely to discover where the boundaries of the unsustainable or sustainable might be. Whereas, then, as we have identified in earlier chapters, much of the recent academic work on sustainability has been dealing more with prematurely finding technical and management type solutions to deep-seated problems. From the point of view more specifically of the food sector, the opposite has largely been the case. Deep-seated and underlying processes have been identified, but these have been too generalised and largely blind to the ways in which they actually implicate environmental changes and unsustainable outcomes. While economic structures and global relationships have been seen as inherently unstable and contradictory, this analysis has rarely extended beyond the strictly economic, social and political bases. Meanwhile, as we indicated

71

above, those interested in sustainability arguments and social 'nature' questions have tended to limit their connection to the broader political economy of food. Political ecology, for instance, has emphasised the deteriorating environmental conditions – the declining material sustainability – of Southern agricultures (see Bryant, 1992).

In the most recent period, with a crisis of the second food regime which holds fiscal, political and environmental causes and the rise of the liberalised globalisation ideology which assumes the efficacy of a re-established 'free-trade' dynamic, it is clear that the question of nature, of the condition of sustainability and unsustainability reach new critical levels. Such tendencies as the uneven liberalisation of agricultural trade, the emergence of regional trading blocs, and the continued reliance of the 'technological treadmill' to 'solve' environmental problems are now more deeply ingrained in the global political economy. Rural sociologists and human geographers, like other scholars in the social sciences are increasingly having to engage with this paradox (sustainability discourse/unsustainable material conditions) of late-modern society. This is not made easier by the albeit tacit agreements between agricultural economists and the political economists, with the former continuing to recognise the benefits of state withdrawal from food markets and the latter still fixated with the critique of the role of the post-war state in creating environmental risk.

Some writers, however, (see Roberts, 1992) are beginning to indicate that the unfolding dialectic between natural and social processes implies both a continual restructuring of the relationship between nature and agricultural production, and a geographical and historically specified pattern to that restructuring. So far this geographical presumption of uneven development as both an outcome and an ingredient of nature–society relations has been largely limited to that of recognition (see Peet and Watts, 1993). Roberts' observations draw attention to the contemporary limits of the political economy approaches of agriculture. The approach has yet to grapple with the ways in which nature is integrated into the uneven development of agriculture. Also, it needs to investigate the ways in which agriculture is part of social nature – part of the modes of social regulation which define and empower the strategies of actors and social formations. For instance, in most settler colonies, and in particular those areas of extensive plantation agriculture (such as the Caribbean), the social regulation of agricultural production was centrally attached to dominant social formations which emerged post-slavery. These dominant formations defined nature in particular ways, which in turn reinforced the mode of social regulation. How these conditions are maintained and then transformed over time and space is a central priority in the study of environmental change as much as it is in terms of understanding the maintenance of capitalist accumulation itself. (What are the agents in the transformation process and how do modes of social regulation evolve?) As our model of the dynamic mode of regulation suggests (chapter 3), one cannot be achieved without the other. By starting to answer these questions, we can begin to develop a broadened and more refined political economy of agriculture and food which makes a space for nature and sustainability questions.

Beyond regimes and systems : the question of viability of modes of social regulation

The workings of international food systems in broad terms tend to suggest that our model of the long run instability in modes of social regulation is always likely to be a factor against which action and particular social formations have to respond (see Chapter 3). Unfortunately, however watertight the regimes of accumulation and social regulation are (as for some considerable time in the cases of the first (extensive) and second (intensive) food regimes at a global scale), it is clear, standing as we do at the end of a tumultuous century, that neither were sufficiently sustainable, either in relational or material terms. Both modes held profound implications for material sustainability; the first: in its assumptions (not that far removed from Marx's notions of historical materialism) about continuing to extend the frontier of agriculture to what seemed to be its infinite extent, and replacing indigenous cultures, wholesale in many cases, by productionist petty commodity producers; the second, in its attempts to intensify production systems and create mass consumer cultures which were completely detached and disconnected from the natural production process itself.

As we have thus discovered to our environmental cost these processes operated in slow and reluctantly changing time periods. Change occurred reluctantly, and only when the mode of social regulation and the particular social formations which it empowers became so illegitimate – economically, socially and politically – that they were forced to collapse. Technical progress and innovation were intensely directed towards upholding rather than clearing away these structures. Materially unsustainable outcomes were continually made legitimate in the name of increasing production for the greater good, or for the more immediate national security of checking balances of payments. Indeed, it is argued by some regime theorists, that it is only the largely undefined and far less coherent third food regime which begins to impute more value to environmental care in food production, supply and consumption. But even here it depends very much which national or regional spaces one is referring to. For instance, the increasing imputation of environmental concerns in Western Europe, through the development of agri-environmental schemes, more 'careful' consumption by consumers, and in a growing concern over the provenance of foods, is based upon an emerging mode of social regulation which still places a strong emphasis upon the constructed distanciatation of environmental risks to distant producers in the South (Arce and Marsden, 1993).

Indeed, one of the key features of the current globalisation ideology which suggests the continued shrinking of global experience, is the ways in which regionally specific modes of social regulation and dominant social formations are capable of creating new distances and boundaries between dominant and dependent spaces and peoples. Again this suggests the need to focus upon where and by whom the boundaries are being drawn, to extend McMichaels' earlier point: how are space and time dimensions formed relationally? Inside the very construction of these relations lie the material aspects of sustainable and unsustainable outcomes.

So far then, despite a substantial amount of work being undertaken on the workings of international food systems, and the evolution of exciting strands of work along a simplified structuralist–social constructionist continuum, the ability to incorporate nature as more than simply a documented pathology to the workings of the accumulation dialectic have remained limited. Where scholars have attempted to go further, for instance in documenting the activities of environmental groups and consumer groups as increasingly salient factors in social change, such endeavours have tended to be marginalised analytically; either as interesting but marginal 'social movements', or as externally derived social institutions beyond the boundaries of the food and agricultural policy and economic community. In short, they have tended to widen the conceptual gaps between structuralist and social constructionist camps. They have rarely been perceived in regulationist or realist terms, as part of the continuing reconstruction of the modes of social regulation, or as part of the basis for maintaining extant social formations.

The body of work does, however, provide some pointers as to how we might overcome some of the analytical obstacles associated with a real incorporation of social nature, or developing a more robust and analytically sensitive understanding of the 'third' food regime. Three such analytical pointers are of relevance to the development of the approach and analysis developed here:

(i)   integrating, through a realist ontology, structural conditions and local material and social mechanisms and outcomes;
(ii)   incorporating centrally an understanding of spatial uneven development in the analysis of food production and supply;
(iii)   developing comparative analyses of the ways in which evolving modes of social regulation condition and sustain the character of agricultural and environmental development.

In progressing these analytical developments we have to focus centrally upon how modes of social regulation become and stay viable; how they are able to accommodate change; and how this change creates the conditions for the realignment of development projects. These development projects may, of course, be more or less materially sustainable. With our interest on the condition of sustainability we have to see these aforementioned features as building blocks. Without their attention and development it will not be possible to achieve sustainable goals. Underlying causes and mechanisms are both relational and spatial. The relations do not occur on the head of a pin. They implicate space. And space fights back and implicates the relations. It is a two-way process.

In the food and agricultural sphere in particular we are reminded by Kautsky (1988) of the inherent spatiality and distinctiveness of the agricultural and food accumulation process. However industrialised and commercialised food commodities become they are largely still subject to natural conditions in their production transfer and consumption. This marks out the agricultural accumulation process from that of other economic sectors. A point recognised only marginally in the literature. In specifically studying the agricultural accumulation process we are always close to nature because it is a means as well as a condition of accumulation and capitalist development itself. It is central to the commoditisation process.

Under conditions when we may wish to consider agriculture and food from a natural standpoint (in assuming a significance to the sustainability discourse, for instance) these distinctive features become particularly analytically important. They provide a basis for situating the specific role of agriculture and food in the global economy as part of nature and of space, as it is of society and polity. The challenge though is to open our analytical frame of reference such that it can incorporate these issues.

Conclusion

Without rendering such an endeavour as 'localistic', 'contingent' or of 'outcome' significance only, it is necessary – by adopting a realist stance – to be more flexible about the transcendence of relational levels in the analysis of food systems. The question becomes not whether we should be interpreting social, economic and political processes at local, national or regional and global scales, but at what level are causal processes and mechanisms constructed by the prevailing and dynamic modes of accumulation and social regulation. What are the appropriate scales at which the legitimisation of the unsustainable (both relational and material) is made, justified and made viable? What are the contradictions, from a realist perspective, in these legitimatory processes? What sort of social formations do they uphold?

These are dynamic and spatially flexible questions to apply to the evolution of agricultural and food systems. And the food system literature, with its focus upon globalisation and wider historical comparative analysis, serves as a useful preface to this. Moreover, it gives an opportunity to consider in our research design and practice ways of focusing upon narrower time and space horizons, while at the same time allowing the analytical transcendence of levels. Adopting realist principles we can begin to apply a more flexible regulationist analysis to particular agrarian spaces and time periods; and to do this, comparatively, in ways which add value to the higher level political–economic analysis. At the same time, by addressing the extent to which material unsustainable outcomes are inherent in the transformations of 'more local' modes of social regulation, it is possible to fully incorporate the natural with the social as a more 'total' form of social mobilisation of agrarian space.

Through these types of analytical postulates it may be possible to steer a course which leads to a much more fruitful and socially comparative analysis of international food systems. And one which at least demonstrates that sustainability and unsustainability have been there all the time. It is just that much of capital accumulation (in this case of the food sector) and its critical analysis in rural and agrarian studies has tended, up to recent times, to largely ignore or at least not accommodate its presence.

5

SUGAR

Introduction

This chapter provides a context for the two case studies which follow by outlining the nature of cane sugar production and the global sugar economy. The chapter begins with a brief description of the characteristics and history of sugar production and moves on to consider current patterns of production and consumption. After outlining key features of the global sugar economy, including the ACP Sugar Protocol of the Lomé Convention, the chapter relates the development of the sugar sector to the theoretical perspectives described in chapter 4.

Sugar

With significant production in over 115 countries, sugar is one of the most widely produced agricultural commodities in the world. In 1993, the world produced almost 111 million tonnes of sugar. At average 1993 prices, the total value of this production was in excess of US$28 billion. Approximately 30 per cent of total world sugar production, worth around US$8.5 billion, is traded internationally each year. Sugar is also one of the world's most widely consumed foodstuffs with mean global per capita consumption amounting to around 20 kilograms per year (International Sugar Organisation, 1994).

Sugar production is closely associated with a range of practices and events which might well be considered to be unsustainable. The negative effects of the Florida sugar industry on the Everglades is perhaps the best-known example of large scale environmental impact (see, for example, Usborne, 1994). In practice, however, whilst these may vary in their scale and significance, most if not all sugar industries involve environmental impacts of some kind (Blume, 1985; Watts, 1987; Abbott, 1990). Links between sugar production and socially and morally unsustainable practices are also well documented (see, for example, Sánchez, 1964; Adamson, 1972; Coote, 1987; Hannah, 1989; Beckles, 1990; Tomich, 1990; Graves, 1993).

Sugar is a compound of carbon, hydrogen and oxygen. It occurs in a number of different forms, but almost all refined sugar is derived from sucrose. Although sucrose is found in all green plants, just two plants are commercially important: sugar cane which grows in tropical and sub-tropical areas; and sugar beet which is produced in temperate areas. Sugar cane production is generally, but not exclusively, associated with less developed countries whilst beet production is essentially a feature of the developed countries of Europe and North America. Despite the dissimilarity of production methods, the sugar produced is essentially a uniform and homogenous product which does not differ significantly in its nature or quality. Thus sugar represents a good example of a major agricultural commodity where the developed and developing worlds are in more or less direct competition with one another.

> The industry is characterised by a number of contrasts and dichotomies which have enabled it to develop along two separate and independent geographical and political lines. Virtually all the world's supply of sugar beet is produced by the developed countries. . . . Cane sugar, on the other hand, is produced principally by the developing countries of Asia, Africa and Latin America.
>
> (Abbott, 1990:1)

Sugar cane

Sugar cane, which belongs to the same plant group as maize and sorghum, is a perennial giant grass which thrives in tropical and semi-tropical climates. All varieties currently used in commercial sugar cane agriculture are hybrids of *Saccharum officianum* L. (Blume, 1985:37). The mature sugar cane plant is 12 to 15 feet high and has stalks of about two inches in diameter. Commercially produced sugar cane is normally grown vegetatively by planting setts – small sections of cane. Each sett grows a stool or cluster of about eight to 12 cane stalks. Cane planted in this way takes about 18 months to mature before a first crop can be harvested. Until the cane plants become established, fields need to be weeded periodically. Within a period of a few months, however, the ground coverage is generally dense enough that weed growth is minimal. After cutting the remaining stubble will sprout again and can be harvested the following year. This practice is known as 'ratooning' and can be repeated for a number of years before the crop needs to be cleared and replanted. Ratooning is profitable because cultivation costs are minimised, but yields and sucrose content fall progressively with each ratoon. Thus an optimal ratoon cycle length, usually about four years, reflects a balance between reduced costs and falling returns (Blume, 1985:75).

Sugar cane will grow in most tropical and sub-tropical climates and in a range of different soil types. It is, however, susceptible to frost damage and requires a particular seasonality in the climate to grow well. In particular, it needs adequate water supplies during certain periods of the growing cycle. In a number of climates sugar cane can be grown well using natural rainfall, although this implies a degree of risk as shortages of rainfall during the growing season can dramatically affect both yields and sugar content. It is not uncommon for irrigation to be used, as it is for example in much of Southern Queensland.

Sugar cane is normally produced monoculturally, although in some countries rotational crops are traditionally included in the production cycle. In part at least because of these monocultural production techniques, commercially produced sugar cane is very vulnerable to a range of diseases, such as smut, Fiji disease and ratoon stunting disease (Blume, 1985; Abbott, 1990). Accordingly, the productivity of the cane varieties used invariably tends to fall within a relatively small number of years and new disease resistant varieties have to be adopted. Even with the constant development of new disease resistant varieties, modern sugar cane agriculture requires large inputs of both pesticides and fertilisers.

Historically, sugar cane production was a highly labour intensive operation often involving extremely unpleasant and arduous work – hence sugar cane's long association with slavery and other forms of coerced labour. A series of technological developments in agronomy, transport and processing have largely transformed the nature of cane production dispensing with the need for large amounts of labour. Traditionally, cane was cut by hand with a skilled labourer being able to cut around one acre per day. Mechanical cane harvesters were developed in the post-war period mainly in Australia. A typical modern cane harvester can cut several tens of acres per day but costs in excess of US$300,000. These harvesters normally cut the cane two rows at a time, separate the trash and chop the cane stalks into short sections known as billets. These are then loaded directly into trailers for transport to the mill. Although a small amount of cane juice is lost when cane is billeted, chopped cane is much more easily transported and processed than whole stalk cane.

The increased use of mechanical harvesters has greatly expanded the practice of cane burning. Cane burning involves setting fire to standing fields of cane prior to harvesting. This facilitates harvesting as it removes the outer leaves of the cane plants. Most mechanical harvesters are designed to cut only burnt cane, although machines are now available which will cut 'green' or unburnt cane. Hand cutters also prefer to cut burnt cane because this saves them the job of removing the leaves or trash by hand after the cane stalk has been cut. Green cane often has a number of agronomic advantages: the trash is ultimately returned to the soil increasing its organic content; and in areas of low rainfall a trash blanket helps to preserve soil moisture. In high rainfall areas, however, green cane harvesting is not always possible as the trash blanket can contribute to water-logging and can lower soil temperatures and thereby impede plant growth.

In some cane producing areas such as Australia the whole of the production process is now highly mechanised and cane is produced with relatively little labour. Conversely, other producers, such as Guyana, remain dependent on the exploitation of large numbers of very poorly paid workers. Even where cane is still cut by hand however, it is now normal for other aspects of the production process, such as loading into trailers or lorries, to be effected mechanically. The levels of mechanisation occurring in different parts of the world are discussed in some detail by Abbott (1990) and Blume (1985).

Milling and refining

Sugar production is an essentially agro-industrial activity – unprocessed sugar cane, or for that matter sugar beet, has little direct utility or value. Turning cane into sugar is normally a two-stage process. The cane is first milled into raw sugar which is subsequently refined into granulated white sugar and other marketable products.

Sugar cane deteriorates quickly when cut and must be processed within a short time of harvesting, usually within a maximum of about fourteen hours. Thus cane growers need to be located within a relatively short distance of the factory which will process their cane. Processing involves washing, shredding and crushing the cane. The cane is then soaked in water which is heated and evaporated to leave sucrose which subsequently crystallises to form 'raw sugar'. This process creates two principle by-products: bagasse and molasses. Bagasse is the residual cane fibre and is used mainly as a fuel to power the processing factories. Molasses is used in the production of rum, cattle feed and yeast, and can be further processed for human consumption. Historically, cane was most usually processed in small mills on the plantations where it was produced. Developments in technology and the need to achieve economies of scale have meant that processing is now almost always undertaken in relatively large centralised factories.

The raw sugar produced in areas where cane is grown is refined into the product familiar to consumers throughout the world in a separate process which normally takes place in the countries where it is consumed. Accordingly, international trade normally involves raw sugar, which is refined into a marketable product in the importing country. Barbados, for example, exports raw sugar to Europe, and this is subsequently refined by Tate and Lyle in London. As much of the value adding is achieved in the refining and marketing of sugar, this arrangement tends to disadvantage producer countries. Historically, there were valid technical reasons why this pattern was necessary. These are now largely redundant as the technology exists to bulk transport refined sugar without any significant loss of quality.

The history of sugar cane production

The history of sugar cane is closely linked to that of the European colonisation of tropical and sub-tropical parts of the world. Historically the vast majority of production was organised on plantation based systems using slavery and other forms of coerced labour. Various legacies of the industry's often unfortunate history remain important today.

According to Blume (1985:30), three phases can be identified in the diffusion of sugar cane:

(1) Fifteenth and sixteenth centuries: dispersal within the American tropics; the colonial plantation based on slave labour developed there.
(2) Nineteenth and early twentieth centuries: diffusion of commercial sugar cane agriculture occurred elsewhere in the tropics whilst still under colonial rule; the plantation system changed in many ways.
(3) After 1950: in the era of decolonisation sugar cane agriculture has been introduced to many, mostly African countries eager to establish a sugar industry. Again, new types of structural systems in sugar cane production developed.

Although amongst the most widely produced and consumed of agricultural products in the world today, sugar has only been consumed in significant quantities since the mid-nineteenth century. Sugar cane was probably first cultivated in South East Asia some 10,000 years ago and spread through South Asia to the Middle East. The earliest references to sugar making appear in Sanskrit literature of the fourth century BC, but sucrose was practically unknown in Northern Europe until around 1000 AD (see, for example, Deerr, 1949; Blume, 1985 and Mintz, 1985). Until the mid-nineteenth century sugar remained an extremely expensive luxury good in Europe, more properly thought of as a spice than a food *per se*. As Mintz puts it:

> in 1,000 AD, few Europeans knew of the existence of sucrose, or cane sugar. But soon afterwards they learned about it; by 1650, in England the nobility and the wealthy had become inveterate sugar eaters, and sugar figured in their medicine, literary imagery, and displays of rank. By no later than 1800, sugar had become a necessity – albeit a costly and rare one – in the diet of every English person; by 1900, it was supplying nearly one-fifth of the calories of the English diet.
>
> (Mintz, 1985:5)

Sugar's transition from a scarce luxury item to one of the world's foremost mass consumption commodities was certainly timely. During the second half of the nineteenth century two developments threatened to undermine the value of what had formally been a rare and expensive commodity. Just as improvements in cultivation and processing technology predicated a changing scale of production for colonially based cane sugar production, Europe began to develop an indigenous beet sugar industry. Mintz sees this transformation very much as part of a larger picture:

> As the first exotic luxury transformed into a proletarian necessity, sugar was among the first imports to take on a new and different political and military importance to the broadening capitalist classes in the metropolis – different, that is, from gold, ivory, silk and other durable luxuries. Whereas the plantations were long viewed as sources of profit through direct capital transfers for reinvestment at home, or through the absorption of finished goods from home, the hypothesis offered here is that sugar and other drug foods, by provisioning, sating – and, indeed drugging – farm and factory workers, sharply reduced the overall cost of creating and reproducing the metropolitan proletariat.
>
> (Mintz, 1985:180)

However, whilst the transformations which allowed sugar production to be sustained and indeed expanded during the late nineteenth century were certainly opportune from the perspective of colonial sugar producers, they were hardly part of some grand and objectively promoted strategy to ensure the future viability of the industry. As Mintz suggests:

> The profound changes in dietary and consumption patterns in eighteenth- and nineteenth-century Europe were not random or fortuitous, but the direct consequences of the same momentum that created a world economy, shaping the asymmetrical relationships between the metropolitan centres and their colonies and satellites, and the tremendous productive and distributive apparatuses, both technical and human, of modern capitalism. But this is not to say that these changes were intended, or that their ancillary consequences were well understood. The ways in which the English became the biggest sugar consumers in the world; the relationships between the colonial loci of sugar production and the metropolitan locus of its refining and consumption; the connections between sugar and slavery and the slave trade . . . these and many other aspects of sugar's history must not be thrown together and labelled 'causes' or 'consequences' as if, once enumerated, they explained everything or anything by themselves. But it is possible to point to certain long-term trends the general consequences of which are readily discerned.
>
> (Mintz, 1985:158)

By the beginning of the twentieth century many of the key features of present day sugar production had already been established. Sugar had emerged as a widely produced agro-industrial commodity. And the duality of beet in the North and cane in the South had already become clear. During the twentieth century, most sugar cane producing countries were to become independent, but sugar's incorporation within a global capitalist economy and the unequal relationships embodied within this have remained highly significant.

Current structure of world sugar production and consumption

Today sugar, either cane or beet, is grown in more countries than any other agricultural product. Overall global sugar production rose from 52 million tonnes in 1960 to around 101 million tonnes in 1982 and continued to expand, albeit slowly, throughout the 1980s, reaching a level over 110 million tonnes in the early 1990s. Within this, developing countries, essentially sugar cane producers, increased their share of the overall total from 52 per cent in 1960 to 58 per cent in 1983. Production in developed countries, primarily of beet sugar, fell from 48 per cent of the overall total to 42 per cent during the same period (FAO, 1987:7). (See table 5.1.)

Although, as Mintz (1985) has observed, developing a cane sugar industry has often proved to be 'like holding Confederate currency', a significant number of Southern countries have either expanded existing production or established new sugar cane industries during recent decades. Several factors appear to underpin this development. These include: objectives of self-sufficiency in sugar production; inappropriate responses to short-lived hikes in the sugar price; inaccurate predictions of future demand; and the ready availability of capital from intergovernmental lending agencies (FAO, 1987:4). One of the most striking examples of a rapidly expanding sugar industry in the South is Thailand. Thai sugar production rose from 676,000 tonnes in 1961–62, to 1.7 million tons in 1981–82 and has continued to expand since then.

The world-wide average consumption of sugar in 1985 was 20.3 kilograms

Table 5.1 Sugar production: developed and developing countries

	1960	1965	1970	1975	1980	1983
Developed countries						
Million tonnes	25	30	32	35	38	40
Percentage	48	47	45	44	46	42
Developing countries						
Million tonnes	27	34	40	45	45	56
Percentage	52	53	55	56	54	58

Source: FAO 1987

per person. However, this gross figure masks large variations between countries. For example, in Australia per capita consumption in 1985 averaged 48.5 kilograms whilst Kampucheans only consumed an average of 0.7 kilograms in that year (FAO, 1987:5). Overall global sugar consumption rose from 49 million tons in 1960 to 95 million tons in 1984, but has been rising only very slowly since then (FAO, 1987; F. O. Licht, 1993). Three important factors emerge from the analysis of recent sugar consumption trends. First, a distinct dichotomy exists between the consumption trends existing in developed countries and those in developing countries. Second, there is a move away from the direct consumption of sugar which is in part being offset by increased consumption in manufactured food products. Third, traditional uses for sugar are increasingly being threatened by the adoption of a range of non-sugar sweeteners.

Direct consumption of sugar is currently falling in most developed countries, largely because of health concerns and the belief that sugars are fattening. Conversely, the demand for sugar is increasing in most developing countries (see table 5.2). Direct consumption of refined sugar is, at best, static in most developed countries. To some extent, this trend has been counteracted by the increased use of sugar in manufactured food products. However, this market is itself tenuous as other major industrial sugar users, such as soft drink manufacturers, have been moving to artificial sweeteners (Heismann, 1993).

The development of various forms of artificial sweeteners has represented a major challenge to the world sugar economy over recent decades. Chemical sweeteners such as saccharine and newer products such as aspartame have now been marketed for some time. Sugar consumption is also threatened by the expanded production of another agro-industrially produced sweetener – high fructose corn syrup (HFCS). The penetration of what were traditionally sugar markets by these sweeteners has at times been quite dramatic. For example, while US sugar consumption fell by 2.2 million tons between 1980

Table 5.2 Sugar consumption: developed and developing countries

	1960	1965	1970	1975	1980	1983
Developed countries						
Million tonnes	32	39	42	41	46	46
Percentage	67	66	60	54	53	50
Per capita (kg)	32	38	40	39	40	38
Developing countries						
Million tonnes	16	20	29	35	41	46
Percentage	33	34	40	46	47	50
Per capita (kg)	8	9	11	12	12	13

Source: FAO 1987

and 1984, consumption of HFCS increased by 2.3 million tons during the same period (Abbott, 1990:333). That said, the potential of these alternatives is itself limited. Whilst their growth has been promoted somewhat by health concerns over sugar, similar concerns are apparent with respect to most chemical sweeteners. Moreover, sugar has a range of organic properties which support its continued use in many products, for example it provides texture and acts as a preservative in manufactured foods.

Special sugars and non-food uses for sugar

Historically, the major by-products of sugar production have been molasses and, derived from this, rum. Molasses is a marketable product in its own right, both for human consumption and in animal foods. Rum, however, is considerably more significant in terms of income generation and the majority of established cane producing areas have rum industries. A more recently developed non-food use for sugar has been in the production of ethanol for use as a vehicle fuel, most notably in Brazil (see, for example, World Bank, 1980). Even given the atypical scale of the Brazilian sugar industry, however, ethanol production has hardly been an economically viable proposition and there seems to be little prospect of this end-use being developed in other producer countries in either the short or medium terms.

There is also some apparent potential in the by-products of cane sugar production. Bagasse, which is the plant material remaining after the cane juice has been extracted, was traditionally burnt to fuel the crushing mills. It can, however, be made into a variety of paper and wood substitute products. A growing interest has been developing regarding the potential of sugar as a feedstuff for various processes within the chemical industry. Whilst this latter possibility may well prove to be very significant in the longer term – non-food uses for sugar and its by-products are likely to remain a minor consideration in the immediate future.

In the short term, at least, the consumption of sugar as a foodstuff, whether directly or indirectly, is the only variable likely to have any truly significant effect on overall global demand. Estimates vary as to how much future growth is likely. Whilst there is clearly potential for increased consumption in the South as intakes approach those in the North and as a result of population growth, a whole range of uncertainties prevent any reliable prediction. Within this, however, it seems unlikely that any future growth will be anything other than modest. Certainly, it seems unlikely that the structural overproduction which currently exists will be negated by any demand side fix. Recent patterns of sugar production and consumption indicate a pattern of distinct and persistent overproduction. This appears to be underpinned by a range of factors. On the one hand, short-lived price hikes periodically result in new and expanded production. Beyond this, the global sugar economy involving as it does a high degree of protectionism and support for domestic industries also means that production is highly price insensitive. A factor which is exacerbated not only by the existence of fixed assets in established industries but also by the relatively long length of sugar cane production cycles (see, for example, World Bank, 1986).

Structural overproduction such as exists within the global sugar economy clearly represents an inefficient use of resources. And whilst this imbalance may well advantage some groups – essentially sugar importers and sugar multinationals – the depressing effects which it has on sugar prices are always going to be likely to promote the over-exploitation of both human and natural capital. In practice, sub-economic sugar industries tend not to be closed down. Rather, they struggle to remain sustainable by adopting more and more exploitative practices. In this sense, at least, a link may well exist between the nature of the global sugar economy and a whole range of unsustainable events and practices.

The global sugar economy: conditions of boom and bust

Throughout its history the global sugar economy has always been typified by boom bust cycles engendered by extreme price volatility. This pattern remains just as valid today as it did in the past (see figure 5.1). Recurrent short periods of high prices for sugar within the world market are seen as problematic because they encourage the entry of new producers and existing

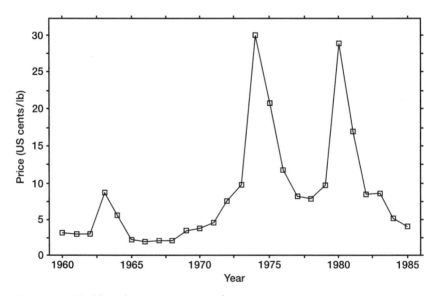

Figure 5.1 World market sugar prices 1960–85
Source: Abbott: 1990

producers to expand production. Borrell and Duncan outline the extent and effects of this extreme price instability in these terms:

> In June 1985 the world market price slumped to an historic low of $0.06 per kilogram. A decade earlier, in the boom year of 1974, sugar had sold for a brief period at around $2.60 per kilogram (in 1985 values) and averaged $1.30 per kilogram throughout the year.
>
> (Borrell and Duncan, 1989:172)

In practice, sugar prices fluctuate markedly not just seasonally but often on a day to day basis. Although this is hardly apparent from figure 5.1, the market price of sugar has been falling in real terms ever since the introduction of sugar beet and the technological transformations of cane production which occurred during the nineteenth century. This underlying trend is disguised by both the volatile nature of the market and inflation but as Mintz (1985:158) suggests, 'the steady and cumulative decline in the relative price of sugars is clear enough'. This decline is significant because in practice it has defined a progressively stressful context within which producers have had to produce sugar more and more 'efficiently' in order to remain competitive and hence sustainable.

International trade in sugar

Although global sugar production has risen in absolute terms the percentage of total sugar production traded internationally has been falling for some time. Over 70 per cent of the world's sugar is now consumed in the countries where it is produced (Abbott, 1990; F. O. Licht, 1993; ISO, 1994). A high proportion of the remainder is exported under 'controlled market agreements' (Coote, 1987:38). In practice, there are two types of controlled market agreement: bilateral agreements; and special arrangements. Bilateral agreements, which accounted for about 15 per cent of sugar exports in the mid-1980s, are normally fixed term contracts between exporter and importer countries which fix the quantity and price of sugar to be traded between the two countries. Brazil, for example, had an agreement of this type to supply the Soviet Union with 320,000 tonnes of raw sugar each year between 1981 and 1985. Australia has entered into a number of such agreements with Japan and other countries. Some 25 per cent of sugar exports occur through special arrangements. During the 1980s, the three most important of these were: the Sugar Protocol of the Lomé Convention, the quota system of the USA, and Cuba's trade with the Soviet Union and other Eastern European countries. This last arrangement is now effectively defunct. Blume summarises the situation in these terms:

> Two sectors can be distinguished in the international sugar trade.

> Some 25 per cent of the international trade is handled under special agreements, such as the 1974 Lomé Convention which regulates the trade between the EEC and ACP countries, and the preferential trade arrangements between Cuba and the COMECON countries. Apart from the sugar exports traded under these special arrangements a substantial amount of sugar is handled under long-term supply contracts, further restricting the world market which as a result is surprisingly small.
>
> (Blume, 1985:301)

With only a small proportion of annual production being freely traded on the open market, the global sugar economy is in practice a 'thin market'. Such markets are very vulnerable to the effects of relatively small variations in output or disruptions to existing trading patterns and tend to react dramatically to any such events. As it currently operates, the global sugar market serves to: (a) increase the volatility in the price of openly traded sugar on the world market; and (b) to depress the price for sugar on the open market in the long term. This last factor is highly significant. Although a significant proportion of sugar exports occur under some form of bilateral arrangement, these are negotiated within the context of structural overproduction and the volatile but normally very depressed prices which occur in the residual market. This clearly prejudices the positions of exporter countries. Even where bilateral agreements are negotiated successfully, these seldom involve any particular price premium over prevailing market prices. Moreover, it is not unheard of for importing countries to default on agreements when market prices fall below those previously negotiated.

Regulation of the international sugar economy

There have been a number of attempts to regulate international sugar trade and prices during the twentieth century. The most significant of these have been those promoted by the International Sugar Organisation (ISO), which incorporates both sugar exporting and importing countries. The ISO has promoted a series of International Sugar Agreements (ISAs). Four ISAs have been instituted since the Second World War: 1953, 1958, 1968, 1977 (see figure 5.2).

These ISAs attempted to keep sugar prices within predetermined bands by allocating Basic Export Tonnages (BET) – effectively voluntary export quotas – to producer countries, and through the development and controlled release of buffer stocks. In practice, the ISAs proved to be almost totally ineffective with prices straying outside bands for almost as much time as they stayed within them during the periods when the ISAs were in operation. When the 1978 ISA lapsed in 1984 it was subsequently extended for a further two years, but it was not possible to negotiate a new agreement. Although the

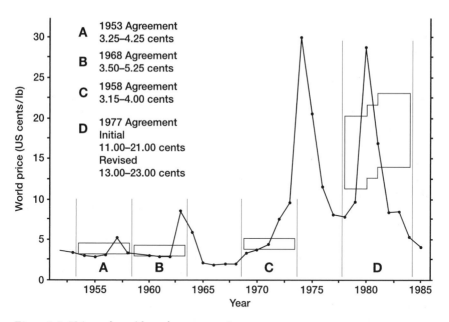

Figure 5.2 ISAs and world market sugar prices
Source: FAO 1985

ISO remains in existence, with offices in London, it now performs a purely administrative role and is mainly concerned with gathering statistics.

The effective failures of successive ISAs reflect various problems which included: non-participation of major parties including at various times both the EU and the USA; non-compliance with BETs; free riding; and demand trends for sugar which have tended to be much more static than has often been predicted (FAO, 1987; Abbott, 1990).

The Sugar Protocol of the Lomé Convention

The absence of any effective overall regulatory framework covering the global sugar economy has ceded a central significance to the policies and practices of the key players within the sugar economy. A small number of multinational companies have become increasingly significant at a global level. Tate and Lyle, for example, a British-based company, had a turnover of £3,817m in 1993. Tate and Lyle either own or have interests in over 90 companies which operate in over 30 countries (Tate and Lyle, 1994). Even more significant, however, are the policies of the major purchasers of sugar on the international market. According to Sturgiss, Tobler and Connell (1988), for example, the joint effects of EU, US and Japanese policies has been to depress the world price by around one third whilst increasing price volatility by 28 per cent.

The USA has various special trading arrangements with sugar producing countries, particularly in Central America, the Caribbean and the Philippines. The rationale for these has often been as much strategic as economic. European Union sugar policy has also had a highly significant effect on the global sugar economy. The EU is now one of the world's largest producers and exporters of sugar. Over and above this, it also has formal trading arrangements with a large number of Southern cane producing countries. These arrangements are formalised under the Sugar Protocol of the Lomé Convention.

The Lomé Convention was first signed in 1975. The objectives of the convention involved granting some protection to 64 African, Caribbean and Pacific (ACP) countries who then had trading arrangements with members of the European Community. The Protocol incorporated those colonies and former colonies which had traditionally exported sugar to Britain under the Commonwealth Sugar Agreement (CSA). Australia was the only CSA signatory which was subsequently excluded from the EU Sugar Protocol. Under the terms of the Protocol, the EEC agreed to import 1.3 million tonnes of raw sugar from the ACP countries (World Bank, 1986:143; Borrell and Duncan, 1989:180). Each sugar producing ACP country was allocated a quota based on historical trading patterns. Barbados, for example, was granted a quota of 54,000 tonnes of sugar. These quotas receive a guaranteed preferential price related to the 'A' quota price paid to European beet producers. Since 1975, European intervention prices have consistently been considerably higher than the world market price for sugar (see figure 5.3 and table 5.3).

Although it was a net importer of sugar in 1975, the EC was exporting over 5 million tonnes onto the world market by 1981 (Coote, 1987:100). Accordingly, all imports from ACP countries have been effectively re-exported onto the world market since the late 1970s. Given the considerable premiums paid to the ACP countries, this has represented a not inconsiderable cost to the EU. However, recent EU Common Agricultural Policy reforms and likely future changes in EU agricultural policy seem to indicate that the future of the ACP agreements are at best uncertain. In practice, the Protocol wording which covers the continuation of the agreement is somewhat ambiguous. Thus whilst the governments of the ACP countries involved tend to argue that the agreement is indefinite, most neutral observers believe its extension beyond the short term is highly unlikely. Any curtailment of the current EU arrangements would, most certainly, have significant economic, social and environmental impacts in many ACP countries.

Sugar – regimes and the conditioning of unsustainability

The dynamics of the global sugar economy, particularly during the last 150 or so years, can be quite readily related to the key strands of academic think-

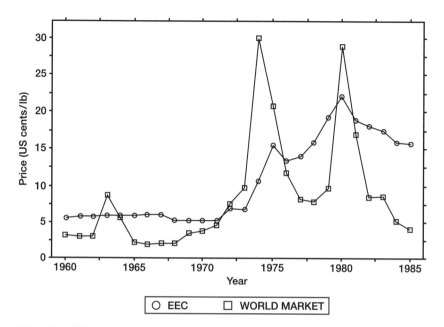

Figure 5.3 EEC and world market sugar prices 1960–85

Source: Abbott (1990)

Table 5.3 Financial benefits of the Sugar Protocol to ACP states (ECUs)*

	1975	*1976*	*1977*	*1978*	*1979*	*1980*	*1981*	*1982*	*1983*
Annual quota (million tons)	1.3	1.3	1.3	1.2	1.2	1.2	1.2	1.2	1.2
Mean world price per ton (i.e. LDP)	273	183	144	143	305	500	280	175	82
EEC guaranteed price per ton	255	296	312	327	339	357	391	401	443
Difference EEC & LDP	18	114	167	184	34	−143	111	235	361
Net gain/loss from protocol	24	148	218	237	44	−184	141	304	467

Source: Abbott (1990)

* Note: 1975 figures in European Units of Account (EUAs)

ing outlined in chapter 4. Certainly, a historical perspective on the sugar sector paints a clear enough picture of a dynamic progressing through extensive and intensive regimes and the possible emergence of a third regime. It also illustrates, quite clearly the inherently crisis prone nature of these regimes.

Sugar was probably the first major food commodity to be effectively integrated within a global capitalist system. By the mid-nineteenth century production, at least, was organised within an extensive regime. The very idea of 'sugar colonies' or 'sugar islands' testifies to this, and institutions such as the CSA and the Sugar Protocol of the Lomé Convention are sensibly understood as measures designed to support such a regime. Equally, it is possible to see a more intensive regime, involving sugar as a mass consumption good and the modernisation of both beet and cane sugar production as having evolved during the second half of the twentieth century. One interesting point here is that the transition from extensive to intensive regimes has been fuzzy and, to some extent, partial. The ACP agreement persists still and modernisation of sugar production has been uneven – quite profound in Australia, for example, and almost non-existent in some Southern regions. And, whilst the present situation is clearly one of transition and uncertainty, as production patterns are changing not least with deregulation in some regions, the precise nature of any third regime remains uncertain.

The sugar sector also demonstrates two other key points. First, it illustrates the uncertain process of conflict and struggle through which new regimes and new modes of social regulation emerge and achieve validity. Consider, for example, the 'fortuitous' shift in consumption patterns which underpinned sugar's place as a mass consumption good and hence extensive production during the late nineteenth and early twentieth centuries or the changing consumption preferences of Northern consumers which have occurred in recent decades (Fine *et al.*, 1996). These were neither planned nor plannable, but still key elements of the development dynamic.

Second, the sugar sector clearly illustrates the ways in which any particular regime and the elements of which it is constituted tend to become increasingly stressed through time. The 'extensive' sugar regime, in its own way, faced the same contradictions and crises as the extensive period of the general capitalist dynamic and within this the first food regime. Similarly, not only did the intensive sugar regime operate under conditions of volatile but generally declining commodity prices, it also generated a range of internal contradictions often mirroring the contradictions of the Fordist regime more generally. Not the least of the problems here being the range of contradictions reflecting the modernisation and intensification of production and the untenable nature of policies providing protectionism and significant levels of price support in the US, Europe and Australia. Moreover, whilst the contradictions inherent in the modernisation and intensification of the sugar sector are numerous and clear enough, the situation in the sugar sector also mirrors the more general situation in that whilst a number of specific environmentally and socially unsustainable outcomes have recently become a focus for concern, they have not as yet been theorised in a way which embeds them convincingly within any metatheoretical perspective.

Individual sugar producing regions are often typified by practices which by almost any definition are variously socially, morally or environmentally unsustainable. We believe that understanding the causality of these unsustainable practices may well provide useful insights into the origins of the unsustainable in the more general case. Albeit for different reasons, the sugar industries of both Barbados and Australia have become increasingly stressed in recent years, and this stress has often been reflected in a range of practices and events which might well be considered to be unsustainable. Not least because the relative transparency of many of the structures and mechanisms which underpin specific examples of unsustainability in the sugar sector is such that a multi-level understanding of this causality may be possible. If this is to be achieved, it is necessary to understand not only the specific, contingent conditions which vary between sugar producing regions, but also the relations between these regions and other parts of the world. It is also important to understand the institutions and values which regulate and serve to reproduce these relationships.

6

THE REGULATION OF THE BARBADOS SUGAR INDUSTRY

Introduction

Barbados is a small, tropical island in the Eastern Caribbean (see figure 6.1). For over 300 years sugar dominated Barbadian development, and almost every aspect of life on the island has been profoundly influenced by this single commodity. However, the Barbadian sugar industry is currently in crisis. This chapter begins with a brief description of Barbados and the island's history. Attention then shifts to the sugar industry and particularly the recent period of crisis. Subsequent sections of the chapter consider how and why the industry has collapsed and how this unsustainability is related to a range of other unsustainable practices and events on the island. What emerges here is a complex picture composed of partial and often contradictory explanations confused by the biases and self-interested perceptions of many of those involved in the industry. Within this, however, it is clear that the crisis cannot be adequately explained by the technical inefficiencies of Barbadian agriculture. Moreover, Barbados' access to protected and highly preferential markets suggests that the current problems cannot be fully accounted for in terms of externally generated pressures. The final sections of the chapter provide a deeper analysis of 'unsustainable events' in and around the Barbadian sugar industry using the approach outlined in chapters 2 and 3. From this perspective, it soon becomes clear that a more meaningful explanation of unsustainability in present day Barbados needs to reflect the unsustainability inherent in the plantation system and the pattern of social relations on the island. The analysis here focuses on the strategies which have been adopted by the island's elite group to sustain their own status and privilege and the institutional and social context which has legitimated and empowered these strategies.

The Barbados model of sugar production: an exercise in constructing sustainability

Both Spanish and Portuguese conquistadores may have landed on Barbados during the early sixteenth century, but it was not until 1627 that English colonists established the first permanent European settlement on the island (Beckles, 1990:7). Although the island had been inhabited prior to this date, there was no indigenous population on the island when these first British colonists arrived. Unlike most other Caribbean islands, which frequently came under the control of first one European power and then another, Barbados was to remain a British colony until independence in 1966.

During the first 20 years of colonisation, experiments with a range of crops including tobacco, cotton and indigo all proved to be less than successful (Watts, 1987). By the early 1640s, however, sugar cane had become established and 'by 1645, Barbadian planters believed that they had found, at last, a truly profitable staple. . . . Sugar cane spread rapidly throughout the island and by the mid-1640s Barbados had emerged as perhaps the most attractive colony in the English New World' (Beckles, 1990:13).

Although sugar could command high prices in Europe during the seventeenth century, the early sugar cane planters on Barbados experienced a range of problems as they strove to develop a sugar industry on the island. Initial attempts were based largely on the 'Pernambuco model' which was already established on the South American mainland (Watts, 1987). The technical expertise gained from Brazil enabled the industry to gain a foothold, but it was some years before Barbadian planters became knowledgeable and experienced enough to adapt the Brazilian techniques to suit local conditions. With the possible exception of moderately inadequate levels of rainfall, especially in the dryer, northern parts of the island, the climate and soils of Barbados are quite well suited to sugar cane production. Nevertheless, sugar cane can be a demanding and difficult crop to grow well. And the initial enthusiasm of early planters was soon to be tempered by the emergence of a range of problems. One concerned the clearing of the island's forest cover which proved to be a labour intensive and expensive procedure. Early planters also experienced a range of agronomic problems including extensive and severe soil erosion and soil nutrient depletion. Although many planters had been taken aback by the totally unexpected fall in yields which accompanied these developments, solutions were soon found for each of these problems. And within a few decades of the first attempt to produce sugar on Barbados an effective and agronomically sustainable sugar production system was already in place.

Several unique practices were developed, such as the specialist dung farms which produced the organic matter necessary for maintaining soil fertility. Another basic but successful and enduring adaptation to the environmental constraints on sugar cane production was the cane hole. These were

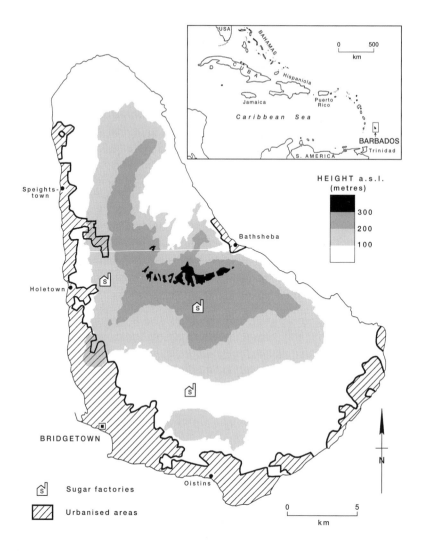

Figure 6.1 Map of Barbados

Barbados is approximately 34 kilometres from north to south, 23 kilometres east to west and has a total area of 430 square kilometres. Although the total population in 1992 amounted to only 259,000, population density is relatively high with 1,677 persons per square kilometre. Population growth is low having averaged considerably less than 0.5 per cent per annum over the last 20 years. Approximately 110,000 people live in Bridgetown, the capital, with the remainder living in either one of several smaller urban settlements or in more rural areas. Life expectancy and infant mortality rates of 73 years and less than 20 per 1,000 live births respectively are comparable with those in many European countries. Mean per capita incomes of US$5,200 in 1991 are relatively high in the context of other Caribbean countries. Health, education and welfare provision are all relatively well developed. The island introduced to the West Indian landscape specifically as an erosion-control measure some time around the start of the eighteenth century. As Watts explains:

> Following the removal of the trash left from the previous crop, a systematic spacing of squares, approximately five feet in size, was marked out by hand hoes. In each square a cane 'hole', which measured two or three feet along each side, then was dug out to a depth of five or six inches. Once excavated, the holes remained unused until they were planted with cane in November: but the very existence of the two-directional system of ridges between them was sufficient to prevent or contain any down slope soil wash which threatened beforehand. . . . Cane holing should not be underestimated as being the first major, reasoned and largely successful attempt at controlling soil loss on sugar estates within the West Indies. Of local origin it was retained for many years.
>
> (Watts, 1987:402).

Whilst the early planters were able to address most of the environmental and agronomic problems which they faced quite effectively, the solutions found often involved highly labour intensive practices. And this in itself was to generate new difficulties as planters consistently experienced problems in ensuring adequate labour supplies.

The planters also faced difficulties in turning cane into a saleable commodity. During the seventeenth and eighteenth centuries, each large estate on the island had its own mill. These processed the cane grown on the plantation, and in some instances that of other smaller producers. There were, however, limits to how much cane could be processed given the technology available at that time, and larger estates would sometimes be broken down into separate production units, each with its own mill. Early mill technology was often

---

has a modern well-equipped hospital and health care is free at source. Literacy rates are comparable to those in the US and Europe. Secondary level education is compulsory and various forms of tertiary education are available. Barbados has a basic but effective welfare system incorporating old age pensions, unemployment benefits and social security provisions. The significance of slavery in Barbadian history is reflected in the fact that well over 90 per cent of the Barbadian population are of African descent. Approximately 4 per cent of the Barbadian population are white, with about 1.4 per cent – around 4,000 – being long established Barbadians. Two thirds of the white population are recent immigrants. Having gained independence in 1966, Barbados is now an independent state within the British Commonwealth. Executive power is vested in the British monarch, represented by a Governor-General. Legislative power is exercised through a bicameral parliament, consisting of an elected 28 member lower house – the House of Assembly; and an appointed 21 member upper house – the Senate. The two main political parties on Barbados are both socialist: the Barbados Labour Party (BLP) and the Democratic Labour Party (DLP).

rudimentary and knowledge of the processes involved were often inadequate. The English were always behind the Spanish in the development and adoption of milling technology, but mill design and techniques were progressively improved over the years with innovations such as the 'Jamaica train' being adopted (Mintz, 1985). Also, animal power was gradually replaced with wind powered mills throughout most of Barbados, which, in their turn, were progressively replaced by steam powered mills, which used the crushed cane stalks or bagasse as fuel.

Once established, sugar cane had spread rapidly through most of the island. Exact production figures for most of the seventeenth century are unavailable, but the island appears to have been regularly exporting around 15,000 tons of sugar per year to England throughout much of the second half of the century (Watts, 1987:285). Although sugar cane came to dominate the island's agriculture, production techniques, particularly milling technology, were such that it was not until the start of the nineteenth century that this level of production was expanded. Various developments during the nineteenth century, particularly the adoption of vacuum pans and centrifuges for processing cane juice into raw sugar, allowed production to be vastly expanded to meet the growing demand for sugar in the new urban industrial centres of Europe. Apart from short periods during the two world wars, when priority was given to food production, sugar output rose steadily until the 1960s when it peaked at around 200,000 tons per year (see figure 6.2).

The nature of early sugar production techniques was highly significant in determining the patterns of development which occurred in Barbados. The practicalities of operating a sugar estate served to determine the pattern of land holdings. In practice, plantations needed to be of a particular size to be viable. This, in turn, determined a particular and enduring pattern of social relations on the island. The costs of establishing and subsequently operating a working plantation – purchasing and clearing the land, building a mill and acquiring and maintaining a large work force – were extremely substantial (Watts, 1987:187). In practice, this served to limit the ownership of plantations to those with access to large amounts of capital or credit. As Watts states:

> A crude, vertically oriented system of land use frequently was put into effect, with the rich bottom-lands being taken over by the 'big' and 'middling' landholders, and the poorer-quality ridgetops and intervening slopes left to the 'smaller' planters and freemen. . . . Allied to this type of planter stratification, which by 1680 had become a fact of life in Barbados, a ruling estate-owner 'aristocracy' or 'plantocracy' also had begun to establish itself, in the sense that a restricted number of families commenced to control island affairs more completely than ever before.
>
> (Watts, 1987:332)

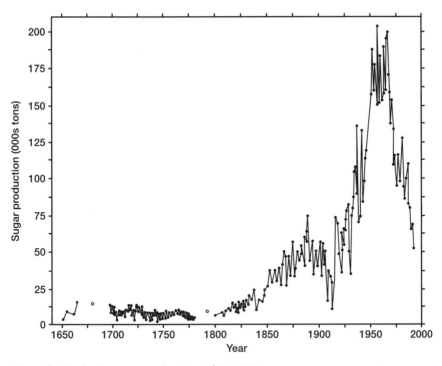

Figure 6.2 Barbados sugar production 1650–1993
Source: Watts (1987)

Unlike the estates in Brazil which commonly operated a share cropping system, whereby landowners leased land to small scale planters in return for a share of their sugar, their counterparts on Barbados tended to work their land themselves. Later technological developments served to reinforce the established pattern of social relations. As sugar mills became more sophisticated only the largest estates could afford to operate such facilities, and thus any smaller cane growers found themselves dependent on larger enterprises for the processing of their cane. As the industry developed, many of the established small landholdings were quickly amalgamated into sugar estates of between 50 and 200 acres in size (Watts, 1987:188). A pattern of land holding which, in its general form, has persisted to this day.

Whereas in the Brazilian model, most estates endeavoured to be as self-sufficient as possible, such a strategy was never pursued on Barbados (Watts, 1987:228). Again this was a feature of Barbadian development which once established was to remain highly significant. Indeed, as sugar estates came to dominate the island, the philosophy of producing a single commodity and sustaining this production by importing all other requirements became a feature of the island as a whole as well as of individual estates.

However, although the early planters in Barbados were successful in developing an effective model of sugar production, and indeed one which allowed sugar production to be sustained for over 300 years, the very success of the enterprise carried with it the seeds of its eventual demise. As Watts puts it:

> Once the raising of cane sugar as a profitable commodity had been ensured, the general tendency was for planters to stay with what they had, rather than indulge in any further agricultural refinement and experimentation. . . . Indeed, one may argue that, in a very real sense, this model was in danger of becoming a fossilised feature of socio-economic life in the Caribbean almost as soon as it had become established, involving as it did structures in its field, factory, social and labour inputs that were so complex that change of any sort was hard to initiate.
>
> (Watts, 1987:383)

Thus whilst the Barbados model of sugar production was sustained over a long time period, the very nature of the model which implied inflexibility and resistance to change was eventually to prove to be its undoing. As we shall see, the eventual unsustainability of the sugar industry owed as much to internally generated tensions and an inertia inherent in its very nature as it did to any externally generated mechanisms.

Equally, whilst the Barbados model of sugar production may have proved to be sustainable in the sense that it existed for a long period of time, it can hardly be seen as being commensurate with notions of sustainable development. Not, at least, if these notions define development as a moral issue. Throughout almost all of its history the Barbadian sugar production has been dependent on the exploitation of various forms of coerced labour. Most significantly, it had a long association with slavery – an association which many commentators suggest continues to be highly significant 150 years after this institution was formally abolished.

Sugar estates on Barbados were first developed through the use of indentured labour, mainly British craftsmen and labourers contracted to work on the island for specific periods. However, it soon became obvious that the scale of labour required could not be adequately met in this way. 'Contemporary opinion in Barbados was that the new sugar estates needed one labourer for every acre of land under cultivation, if all stages of production and milling were to be undertaken with reasonable efficiency: and after the first few years of experimentation with the crop, this requirement was doubled' (Barrett, 1965, quoted in Watts, 1987:202). Accordingly estate owners turned to the purchase of slaves to meet their labour requirements. There were 6,000 slaves working on estates in 1650 and around 20,000 by 1653. By the mid 1660s there were more slaves on the island than there were whites. By 1833, the slave–white ratio had reached over six to one, with over 80,000 slaves and less than 13,000 whites.

Although slavery persisted until emancipation in 1833, a number of tensions had begun to emerge in this form of social regulation well before this time (Watts, 1987:218). The high slave–white ratios heightened fears of slave revolts. The costs of acquiring and maintaining slaves had also risen over the years and many estates experienced difficulties maintaining adequate labour supplies.

Because of the small size and singular development, the post-emancipation situation in Barbados was somewhat different to that in some other sugar colonies, for example Jamaica. Relatively little opportunity existed for slaves to leave plantations and engage in some form of subsistence agriculture. Virtually all of the land on the island belonged to the estates and there were no virgin areas to which freed slaves could migrate. The system was one of increasing rigidities. As Adamson puts it:

> Post-emancipation provides the cliché that everything must change in order that everything must remain the same. . . . The Negro was liberated from the plantation but, he was not free to develop his own economy and culture.
>
> (Adamson, 1972:255)

Sánchez evaluated a similar situation in the neighbouring island of Antigua:

> His Britannic majesty's new subjects learned that the planters had agreed with one another to fix a salary for all the island (i.e., on abolition in Antigua) of 1 shilling for the most skilled workers and 9d for the rest. This wretched wage was less expensive to the planters than the maintenance, clothing, and lodging of each slave. The planters gained from the emancipation above and beyond the indemnity they received from their motherland.
>
> (Sánchez, 1964:24–5).

Thus whilst emancipation may have ended slavery *per se*, the basic patterns of social relations established in Barbados in the seventeenth and eighteenth centuries persisted in all but name until well after 1833. Interestingly, the mode of social regulation remained largely intact. Indeed according to some commentators it still persists:

> The plantation system, the race relations system, the managerial ideology, all these things are a legacy of slavery. Policy in the sugar industry today has to be seen as a survival of attitudes that have survived from the slavery period. Their approach to labour: cheap, cheap wages, lack of sanitary facilities, lack of continuous

employment, these attitudes, as far as I am concerned, have all survived from slavery. Here is an industry that uses 80 per cent of the agricultural land in this country and that land is controlled by the white community that is only 1.4 per cent of the population . . . to my mind that is unjust and needs to be rectified, this country needs land reform which places the ownership of the sugar industry in a larger number of people.

(Personal communication, Barbadian academic)

The sugar industry in post-independence Barbados

Until well after the Second World War the sugar industry remained the dominant economic activity on Barbados. Sugar production which stood at about 50,000 tons in the first decade of the century tripled in the period up to 1970, peaking at a record high level of over 200,000 tons in 1967. As Worrell puts it: 'In 1946 . . . most economic activity depended on overseas trade. Sugar production dominated, accounting for over a third of GDP and bringing in two thirds of receipts from the sale of goods and services abroad' (Worrell, 1982:1). Throughout the post-independence period, however, the sugar industry has become progressively less significant. As early as 1979 an official report into the industry noted the declining significance of this sector in Barbados:

Until about 20 years ago, sugar was unquestionably the mainstay of the Barbadian economy. Since then, the industry's share of gross domestic product has diminished, as a result of both the decline of sugar production and the growth of other industries, principally tourism. From a third of gross domestic product in the mid-1950s, sugar's contribution fell to about 20% in the early 1960s, and in the last three years has hovered around 6%.

(McGregor et al., 1979:44).

In 1991 sugar accounted for only around 3 per cent of GDP, with non-sugar agriculture accounting for a slightly larger figure. By 1992, Barbados' income from sugar exports amounted to only US$33.3m, which represented less than 23 per cent of the total value of domestic exports. The tourist and manufacturing sectors were both considerably more significant than sugar by this time. (See figure 6.3.) Within a period of less than 30 years, Barbados has moved from almost total dependence on a single agricultural commodity – sugar – to a similar dependence on its tourist industry which had developed rapidly in the 1960s and 1970s (Worrell, 1982:8). In 1992, with 432,000 visitors spending US$462m, the tourist industry accounted directly for 16 per cent of the work force and contributed 11.4 per cent of GDP (Pattullo, 1996). Various light industries, particularly electronics, accounted

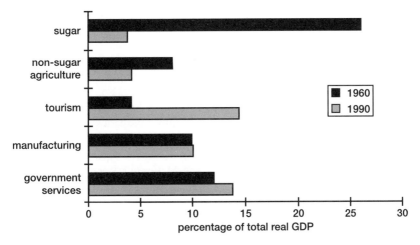

Figure 6.3 Composition of GDP in Barbados 1960 and 1990
Source: Barbados Statistical Service (1992)

for 12 per cent of GDP and 60 per cent of the value of exports in 1992 (Government of Barbados (GOB), 1993).

Not least because of the decline in sugar income, Barbados has consistently experienced deficits in its visible trade balance throughout the last 20 years (GOB, 1988, 1993). This deficit peaked at US$493m in 1990. Although the visible trade deficit is in large part negated by invisible exports, essentially tourism, Barbados' international debt has risen quite dramatically, climbing from US$381m in 1982 to almost US$1,000m in 1992. This latter figure equates to around 25 per cent of GDP and servicing this debt accounted for 17.8 per cent of the value of exports in 1992 (GOB, 1993:39). The inflation rate which stood at over 14 per cent in 1980, fell consistently up until 1986 when it reached a figure of less than 2 per cent. By 1992 it had climbed to around 6 per cent (GOB, 1993:13). Unemployment rates which were around 10 per cent in 1981 averaged around 15 per cent throughout most of the 1980s and stood at 13 per cent in 1992 (GOB, 1993:33).

Prior to Barbados gaining its independence from Britain in 1966, the island's sugar production was sold into three or four main markets. Some 5,000 tons were exported annually to the USA at a premium price under the terms of the US Sugar Act. Around 15,000 tons were used to meet domestic demand. And during years of exceptionally high output Barbados also sold some sugar onto the world market. The principal market, however, was with Britain under the terms of the Commonwealth Sugar Agreement (CSA). Under the terms of the CSA, Barbados was permitted to export some 136,610 tons of sugar per year at premium prices, an arrangement which afforded the island significant benefits. In 1962, for example, prices paid to Barbados under the CSA amounted to B$219.66 per ton whereas the average world market price during that year was B$124.71.

Following Britain's entry into the EEC, the CSA was extended for a number of years, but was subsequently replaced by the Sugar Protocol of the Lomé Convention in 1975. All the former parties to the CSA, with the single exception of Australia, were incorporated within the Protocol. These African, Caribbean and Pacific (ACP) countries, including Barbados, received both a guaranteed market into Europe and preferential prices – equivalent to the 'A' quota price for European beet sugar. Barbados received a quota of 54,000 tonnes of raw sugar per year. As the McGregor Report observed in 1979:

> this price arrangement is more favourable than that of the former Commonwealth Sugar Agreement to the extent that the Lomé price is indexed to EEC prices . . . normally EEC prices are likely to be substantially higher than free world market prices.
>
> (McGregor *et al.*, 1979:30)

Barbados' quota into Europe was substantially less than the one it had enjoyed within the CSA. Given, however, that production was already falling at this time, the majority of any surplus production was still accounted for by the US quota, the Canadian market for molasses and domestic demand, and little if any sugar was traded on the open market.

Perhaps not surprisingly, following a period of high returns during the 1960s, sugar was still seen as being highly significant and quite viable at the start of the 1970s. And although the industry had started to contract, attitudes were still largely positive at the end of the decade (McGregor *et al.*, 1979:11). By 1992 however, cane production in Barbados had fallen to eighteenth century levels. Production fell from a high of over 2 million tonnes of cane (tc) to a low of 528,000 tc in 1967. In 1960, the sugar industry had accounted directly for 20 per cent of the island's labour force, by the early 1990s only just over 2 per cent of the island's population were directly employed by the sugar industry. Sugar's contribution to GDP also fell sharply. As late as 1980 GDP from sugar had amounted to B$94.2m (6.3 per cent of GDP), by 1990, however, sugar accounted for only B$58.5m (2 per cent of GDP) (Sparks Companies Inc. (SCI), 1992:20).

The modern structure of production

Some 100 plantations have remained the principal cane producers in Barbados, and by the early 1990s, they accounted for well over 90 per cent of total sugar production. As figure 6.4 shows, the plantations account for just over 25,000 acres or approximately 80 per cent of the arable land on the island. Most of these plantations have between 100 and 400 acres of cane land, the median size is just below 300 acres (Booker Tate, 1993, vol. 3). The majority of these plantations are privately owned and operated, employing their own labour and using their own equipment. However, as McGregor *et al.* put it in 1979:

> The number of estates belies the degree of ownership concentration. Three companies – two public and one private – together with their subsidiaries own approximately 20 per cent of the cane land. In addition the government-owned Barbados Agricultural Development Corporation (BADC) controls ten estates with nearly 1,300 hectares of cane land.
>
> (McGregor *et al.*, 1979:18)

By the early 1990s land ownership had become even more concentrated with five companies controlling 35 estates which amounted to 15,662 acres or 55 per cent of the harvested area (SCI, 1992:25).

It is common practice on Barbados for sugar estates to be managed by an 'attorney'. In the Barbados context, an attorney is the owner's representative and is given full responsibility for the running of the plantation. Attorneys are usually established and respected planters in their own right who are contracted to run other plantations. It is argued, at least by the attorneys, that there are a number of advantages to this system. In theory, they should, for example, be able to gain significant economies of scale by effectively treating the estates which they control as a single operational unit. Attorneys are paid a fixed fee per acre per month. Historically attorneys were normally employed by British absentee owners who often never visited their estates. In recent years it has become more common for attorneys to be managing estates owned by upper middle class Barbadians or corporate owners who concern themselves with other business interests and delegate the management of their estates to these people. In 1988, some 85 per cent of the sugar estates on Barbados employed an attorney, and just seven attorneys controlled some 63 per cent of the arable acreage on plantations (SCI, 1992:25).

In 1992, the GOB owned nine plantations, with just under 3,000 acres of cane, which it managed through the BADC. Although many of these estates are situated in agronomically marginal areas, government policy has involved maintaining sugar cane production, not least because this was seen as an effective erosion control measure (SCI, 1992:25). In practice, production has been low on these estates and the management of the BADC has been subject to widespread criticism.

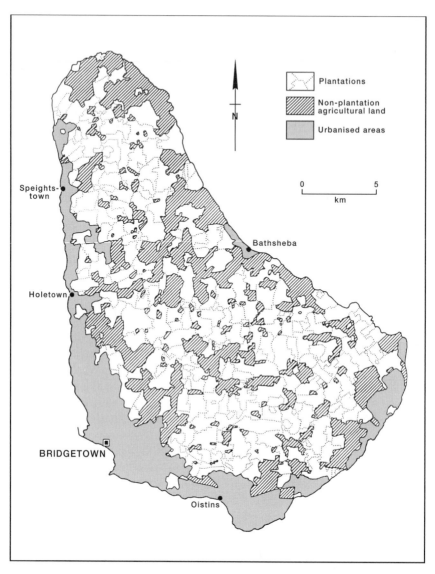

Figure 6.4 Barbados plantations in 1979

Source: McGregor (1979)

Production methods

Barbadian sugar output increased significantly between 1850 and 1960. However, whilst Barbados was able to achieve large increases in sugar output through the use of chemical inputs, new higher yielding cane varieties and larger and more efficient mills, the industry has never been at the forefront of technological development. Indeed, the Barbadian sugar industry of the early 1990s is largely founded upon an unholy mix of mid-twentieth century technology and seventeenth century attitudes. The plantations on Barbados are a legacy of over 300 years of sugar production. Many of their characteristics and many of the production methods used today remain little changed from those which existed in the seventeenth century. These systems are explicitly embedded in history, a situation which is very different to that in countries such as Australia where the mode of social regulation incorporates a strong modernisation ethos.

Fields are often small – optimised to labour intensive production techniques. Typically, plantations are also criss-crossed by numerous cart tracks which have often been deeply eroded into the landscape by over 300 years of use. Although some effort has been made to consolidate and rationalise field patterns, many areas remain unrationalised. To some extent, this may simply reflect the quite substantial costs involved. Equally, however, the conservative attitudes of many landowners are of significance.

Topographical constraints have also acted as a barrier to modernisation. Relatively steep slopes were not problematic when traditional labour intensive production methods were being used, particularly when cane holing was used to prevent soil erosion. Much modern farm machinery, particularly cane harvesters, however, cannot cope with slopes much in excess of 10 degrees. The McGregor Report estimated that about 25 per cent of plantation land was unsuitable for mechanisation. Accordingly, a number of plantations, mainly but not exclusively those in the Scotland District, have proved to be physically unsuitable for mechanised sugar cane production. Moreover, 'the areas potentially mechanizable are unevenly distributed. Few estates on the island are totally mechanizable' (McGregor, 1979:114). Although almost all cane is now loaded and transported mechanically, less than 50 per cent of the cane was actually cut by machine during the early 1990s. This is problematic not simply because it requires high levels of labour, but also because it can result in the harvest not being completed before the onset of the summer rainy season.

During the late nineteenth and twentieth centuries, plantation based mills were gradually replaced by centralised factories. Initially, a relatively large number of factories were operated, but progressive technological development has seen the majority of factories close. Of the 14 which existed in 1970 only three were still operating in 1993: Bulkley, Saint Andrews and Port Vale. All of these factories are relatively small by modern international standards, and operating costs are amongst the world's highest, at around B$500 (US$250) per ton of sugar in 1990 (SCI, 1992:23). For the most part, factory equipment is old, outdated and inefficient. Despite this, capacity in each of the three factories is well in excess of recent production levels. In total, the remaining factories have a capacity of around 800,000 tc per year, which would produce approximately 90,000 tonnes of sugar (ts) (Booker Tate, 1993, vol. 2).

In an attempt to rationalise the milling sector in 1973, Barbados Sugar Industry Limited (BSIL) was formed to operate the sugar factories and to supervise transport and storage of sugar cane, raw sugar and other sugar products. It also had some responsibility for research and co-ordination within the industry. For example, it was responsible for co-ordinating cane harvesting to ensure a smooth throughput of cane in the factories and thus avoid delays during which time the sugar content of the canes would deteriorate. Share ownership in BSIL was restricted to those major land-owners who produce cane. The company was managed through a number of committees and a full-time managing director. The GOB had a seat on the board through which it represented the interests of small farmers and government owned plantations. According to Booker Tate:

> BSIL tends to operate as a co-operative working for the producers. As a result it focuses more on distributing sugar proceeds to growers than generating profits and dividends.
>
> (Booker Tate, 1993, vol.1:3)

Government support for the sugar industry during the 1980s

Although the Barbados sugar industry had already started to contract during the 1970s, the industry was plunged into crisis when currency fluctuations produced a sudden drop in income during 1981. Both planters and the factory sector rapidly experienced liquidity problems. These problems led the Barbadian government to embark on what was, given the small size of Barbados, a massive programme of support for the sugar industry. By the end of the 1980s the total sugar industry debt to the Barbadian government amounted to more than B$1,000 for every person on the island.

Recent government intervention in the Barbadian sugar industry needs to be seen in the context of a long established tradition of such support. Within this, however, a number of specific reasons have been cited to legitimate what was an extremely high level of support. Although earnings from sugar exports had fallen dramatically, sugar exports still represented one of very few sources of foreign exchange for Barbados. It was also widely claimed that unlike the situation with the tourist industry where a high proportion of inputs are imported, most of the earnings from sugar are retained within the island. At this time, direct permanent employment in the sugar industry accounted for about 2 per cent of the island's labour force – and around double this figure during the harvest period. Overall some 4,000 families derived some form of direct income from the industry, and further employment was generated in a range of functionally related activities such as transport. Although agricultural employment is generally unpopular amongst the Barbadian population, a further reduction in employment opportunities would most certainly have been politically negative, particularly in a situation of relatively high unemployment.

Production costs for Barbadian sugar are far higher than the world market price for raw sugar and for some decades Barbados sugar production has only been profitable and viable in the context of the ACP agreement. Moreover, the 63,000 tons produced in 1981 was not sufficient to meet both domestic demand and Barbados' quota into this market. Although the Barbadian government maintains that the ACP protocol is indefinite and cannot be revoked, it is not clear whether this is indeed the case as the protocol wording is ambiguous (Booker Tate, 1993, vol. 1). Certainly, persistent overproduction of sugar within Europe and changed strategic concerns mean that the EU faces pressures to revoke the ACP agreement and thus that it would be unwise for any country to further prejudice its position by fulfilling its quota with sugar purchased on the open market. As the GOB was certainly aware, any continued failure to fulfil their quota, or for that matter continuing to meet domestic demand with imports from the world market as has happened on several occasions, would almost certainly result in the subsequent loss to Barbados of this extremely preferential market.

As explained in Chapter 5, once sugar cane is cut, it needs to be processed within about 14 hours or the sucrose content decreases rapidly. This processing can only be achieved efficiently in centralised factories. Such factories are necessarily of such a size that a minimum volume of cane is needed to support them. Thus, even within the context of Barbados' relatively small factory sector, a 'critical mass' exists below which the factories would be under-utilised to the extent that they could have no chance of operating profitably. Thus in a situation where less cane is being produced, the essentially fixed costs of the milling sector become increasingly significant and will potentially undermine the profitability of the industry as a whole. A similar argument can be seen to exist with respect to other infrastructural and functionally related activities, such as the purpose built bulk sugar terminal used for handling exports.

Despite its rapidly falling acreage, cane remains the dominant land use on the island. In this respect, sugar cane is seen as being important to the island's environment. Certainly many of those involved in the sugar industry argue that it is highly significant in preventing soil erosion. It provides uninterrupted ground cover for a period of at least four years and thus protects the thin and easily eroded soils. In practice, several areas in the

Scotland District, where large areas of sugar cane land have been abandoned or allowed to go to grass, have experienced severe problems of erosion in recent years. During the 1980s, soil loss on vegetated plots in this district averaged 26.1 tons/soil/ha/year, whereas on bare plots it reached rates of over 319 tons/soil/ha/year (Soil Conservation Unit, 1987). That said, the effectiveness of sugar cane in preventing soil erosion remains questionable. Although it may provide more continuous ground cover than many crops, the harvest occurs immediately prior to the rainy season and thus cover is at a minimum when it is most needed.

The nature of sugar cane agriculture is also seen as being significant to the hydrology of the island (Barbados Water Resources Group, 1978). The limestone geology of the island means that there is little surface water on Barbados and most of the island's water demand is met from groundwater sources and springs which occur around the base of the limestone cap. Rainfall infiltrated during the summer rainy season either recharges the aquifer or occurs as throughflow to these springs (Nurse, 1978; Antoine, 1989:4). This throughflow takes several months and accordingly the springs are most productive after the end of the rainy season. Any large scale change in the island's vegetation cover, such as would occur if sugar cane production were to cease, would effect evapotranspiration and infiltration rates and might well impact severely on the established hydrological regime in ways which would prejudice the island's water supply security (Hudson, 1987:17; Trotman, 1994). Barbados' water supply security is also threatened by changes in albedo resulting from widespread land-use change which may increase evaporation rates (Watts, 1997).

The sugar industry is also seen as contributing to the amenity value of the island's landscape. Given the current dominance of tourism in the economy, maintaining a landscape commensurate with tourists' perceptions of a green and productive tropical island is seen as highly significant (SCI, 1992; Booker Tate, 1993). The recent environmental degradation which accompanied the collapse of the sugar industries on other Caribbean islands, such as Antigua (Caribbean Conservation Association, 1991; Government of Antigua, 1991), is frequently cited as an example of the reduction in amenity value which can accompany the rapid collapse of a sugar industry.

A further argument frequently cited by many planters to legitimate support for the sugar industry is that few if any alternative forms of agriculture are viable on the island. This would appear to be a dubious contention. The Barbados Ministry of Agriculture insists that a more diversified agriculture is both desirable and possible and they have maintained this as a primary policy objective for some time (GOB, 1956, 1965). Although non-sugar agriculture is now more significant in terms of GDP, sugar cane still accounts for a very high proportion of all agricultural land on the island. Indeed much of the value of non-sugar agriculture on Barbados is accounted for by the intensive production of chickens and pigs which has developed in recent years. In practice, the reasons why only limited progress has been made in promoting a more diversified agriculture are not straightforward. The market for traditionally produced root crops, such as cassava and sweet potatoes, has been undermined by changing eating preferences. The local market for agricultural products is small and much of this is accounted for by the tourist industry which demands consistently high quality and security of supply. Also, export markets for high value agricultural commodities which might be appropriate to Barbadian conditions, such as cut flowers, are already highly competitive. But perhaps more significant than any of these factors, is the fact that Barbadian agriculture is effectively controlled by a very small group of people whose perceived self-interests may well not be suited by the development of a more diversified agriculture on the island. Moreover, although the absolute and relative significance of sugar in terms of income and employment generation has declined appreciably over the last few decades, sugar and related issues remain highly politicised on Barbados and problems in the sugar sector reflect badly on the GOB.

It is also argued by both the government and those involved in sugar production that further investment in the Barbados sugar industry is rational because the large amounts of investment already committed to the industry cannot be recovered. Thus existing investments should be discounted in any cost-benefit analysis. In practice, this is a self-reinforcing argument as the case for further support increases each time more investment is made. However, even if a policy decision was made not to support the Barbados sugar industry, there would still be a need for careful management if a number of problems were to be avoided. A rapid and total collapse of the sugar industry would almost certainly leave a vacuum which would, in all probability, produce a range of negative economic, environmental, social and political consequences. That said, any planned and phased closure of the sugar industry would be difficult for technical reasons, not least because of the 'critical mass' problem. Notwithstanding these problems, managed retreat would seem to be preferable to a laissez-faire approach which simply allowed the industry to collapse. Thus there may well have been a strong case for short-term support of the industry, for example, whilst new forms of diversified agriculture were established. In practice, however, the nature of GOB support for the industry has extended beyond this.

Support mechanisms

In practice, the majority of GOB support provided to the sugar industry during the 1980s was furnished through the Barbados National Bank (BNB). The BNB is a parastatal organisation. It was incorporated in 1978 as an amalgamation of several government owned financial institutions, including the Sugar Industry Agricultural Bank. The BNB's assets amounted to B\$526.4m in 1989 (SCI, 1992:55). The bank's remit is technically wider than the sugar industry, including, for example, a role in the development of social housing, but in practice its portfolio is dominated by loans to this sector. Although it claims to operate as a commercial bank, the status of the BNB is somewhat ambiguous. Its intended objectives are broadly those of a development bank. Indeed, it legitimates its lending, including loans to the sugar sector, in terms of its developmental function.

The BNB provided soft loans to the sugar industry throughout the 1980s. Interest rates were capped at a maximum of 8 per cent, although normal interest rates were well into double figures for most of this period. These loans were granted with little if any regard for normal banking criteria: income generation potential, equity or whatever. By the early 1990s, the total industry debt to the BNB was in the region of B$250m. Senior managers within the BNB maintain that lending policy throughout the 1980s was subject to direct political control. This is a credible contention. The GOB is the only shareholder in the bank, the board of the BNB consists solely of political appointees and any substantial loans have to be authorised by the finance minister. The GOB also supports the sugar industry through a system of controls on the domestic price of sugar which produced B$14m in income for the industry in 1992 (Booker Tate, 1993, vol. 1:8).

Throughout the 1980s, the milling sector received considerable support from the GOB. Both the operational and strategic capital requirements of BSIL were financed almost exclusively by the GOB through the BNB. Loans to BSIL during this period exceeded B$100m (Booker Tate, 1993, vol. 1:6). Further to this, the GOB guaranteed sugar industry bonds issued by BSIL in the early 1980s to a total value of B$36m.

Traditionally, the plantations had been financed through a number of private banks. However, the commercial banks effectively ceased to do business with the sugar industry in Barbados in the early 1980s. Commercial bank credit to the whole of the agricultural sector averaged less than B$30m – about 3 per cent of total lending – during the 1980s, and little if any of this credit applied to sugar cane production (GOB, 1993:42). During the 1980s the BNB provided the sugar sector with credit some eight times higher than the total extended by commercial banks to the entire agricultural sector (GOB, 1993:24). Whether it was because the commercial banks declined to provide credit or because they chose to use the services of the BNB, most planters sought to finance both long- and short-term capital requirements through the BNB (Booker Tate, 1993, vol. 1:4). In 1992, the BNB's loans to the agricultural sector totalled B$186.6m, with about 90 per cent of this figure comprising loans to sugar plantations (SCI, 1992:55). By the end of the 1980s, GOB loans to the plantation sector amounted to B$6,800 (US$3,400) per acre of estate land on the island – a figure considerably above the value of the land.

The Barbados sugar industry in the early 1990s

Sugar production

Support for the Barbados sugar industry, directly by the GOB, and indirectly by the EU, amounts to the maintenance of an industry in stasis. It has protected the industry from the rigours of globalisation – a situation in stark contrast to that in countries such as Australia. Moreover, despite massive government support during the 1980s, it was unavoidably apparent to all concerned that the situation of the Barbados sugar industry had gone from bad to worse during the 1980s. Sugar production which had already declined by over 60 per cent since 1967, fell by a another third between 1981 and 1992, dropping from 75,000 tons to around 50,000 tons. This drop in output reflected both a reduced acreage under sugar and falling yields (see figures 6.5 and 6.6). In 1967, cane was harvested from 52,000 acres at an average yield of 35 tc/a. In 1992, 22,000 acres were harvested with an average yield of 24 tc/a. (McGregor *et al.*, 1979:44; Booker Tate 1993, vol. 1:4).

By 1992, the US quota had not been fulfilled for some years and the

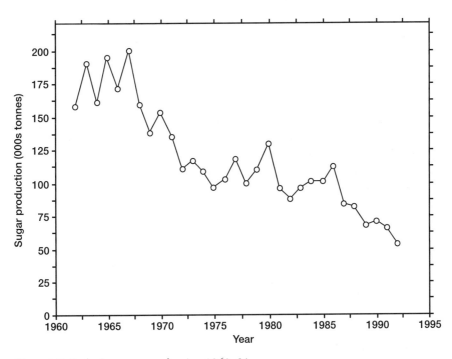

Figure 6.5 Barbados sugar production 1962–91

Source: BSIL Records

113

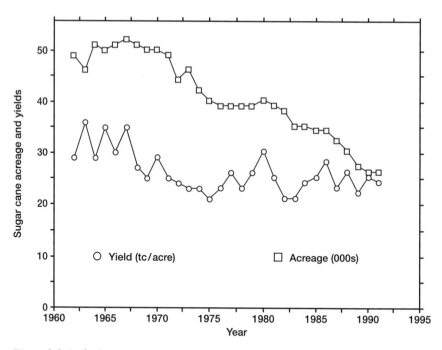

Figure 6.6 Barbados sugar cane acreage and yields 1962–91
Source: BSIL Records

profitable molasses trade with Canada had been lost. Even more significantly, production in 1991 was less than the ACP quota (54,000 tonnes) to the EU. The Barbados government maintains that it was able to meet the 1991 quota by using buffer stocks, although it does concede that sugar was imported to meet domestic demand (personal communication, Chairman Barbados Agricultural Management Corporation (BAMC)). Whilst it may be relatively apparent why this claim should be made, its validity is nevertheless somewhat questionable.

The financial crisis of the sugar industry

By the early 1990s, it was clear that the support afforded to the industry by the GOB during the 1980s proved to be largely ineffective, and moreover, that this level of support could not be sustained. Both the factory sector and many individual plantations had accrued debts which they had no hope of servicing. And at a macro-economic level, the overall industry debt was beginning to have serious consequences for the GOB.

In 1992, BSIL had liabilities of B\$170m and assets of B\$40m. The majority of this debt was owed to the GOB although loans were also outstanding to the Caribbean Development Bank and to a range of private creditors. BSIL received loans throughout the 1980s to finance both capital expenditure and their day to day operations. However, despite the support which BSIL had received, the mills remained outdated and inefficient by international standards. Some investment had been made on upgrading machinery, particularly at the Port Vale factory, but despite this BSIL still experienced operational losses every year throughout the 1980s. By the early 1990s, the company was clearly unable to operate profitably or to service its outstanding debts. Although Booker Tate estimated BSIL's assets at B$40m, even this low figure needs to be treated with some circumspection. In practice, BSIL assets, which essentially consist of three sugar mills, only have any value if a viable sugar industry remains on the island. If the industry were to close, it is unlikely that BSIL's assets would have any real value, certainly little if any of their mostly old and outdated plant could be sold abroad (Booker Tate 1993, vol. 1:6).

The indebtedness of plantations to the BNB at 30 June 1992 was B$113m. Approximately B$30m was owed to other creditors. Some 52 plantations were unable to service their debts to the BNB at this time. Of these 40 were categorised as Heavily Indebted Plantations (HIPs) and 12 as Moderately Indebted Plantations (MIPs). The distinction being that HIPs were not in a position to continue production because they could not meet their day to day operational costs. HIPs accounted for 13,000 acres or 46 per cent of the cane land in Barbados (Booker Tate, 1993, vol. 1:6). Whilst the BNB's accounting procedures effectively separate the agricultural and milling sectors of the sugar industry, in practice, the debtors are the same people. As share ownership in BSIL was restricted to estate owners and virtually all planters are members of BSIL, both sets of debt are owed by the same group of companies and individuals.

A total of B$249m (88 per cent) of the total industry debt could be regarded as delinquent in June 1992 (Booker Tate, 1992, vol. 1:6). To be properly understood, this level of debt needs to be considered in the context of the small size of Barbados and its economy. Of the 100 plantations on Barbados 52 were categorised as being either HIP or MIP. Individual HIPs typically had debts of around B$2.5m. Of those 48 plantations not classified as either HIP or MIP, ten were already owned and managed by the Barbadian government (AIMS, 1991).

By the early 1990s, two things were quite clear. First, it had become unavoidably obvious that despite the very high level of support which it had received, the sugar industry was in crisis. Second, it was equally apparent that this level of support could not be maintained. A large number of growers could not operate because of their indebtedness, and more and more land was being taken out of sugar production. Production in 1991 had not been sufficient to meet both domestic demand and the 54,000 tonnes ACP quota into the EU. Sugar was being imported from the world market, ostensibly to meet domestic requirements, and the continuance of the quota arrangements was becoming increasingly prejudiced. The level of sugar industry debt to the BNB was such that it was placing considerable pressure on the Barbados currency and the IMF were keen to see positive action taken. In fact, the situation was such that unless radical measures were taken very quickly the industry would, in all probability, have ceased to exist within the next year or two.

The Barbados sugar industry restructuring plan

In 1992, tenders were invited from firms wishing to plan and manage a restructuring of the industry. Three firms submitted tenders, and in 1993 Booker Tate, a jointly owned subsidiary of Tate and Lyle and Booker plc were commissioned to formulate and manage a restructuring programme. It is somewhat unclear to what extent the engagement of a foreign firm to manage the sugar industry restructuring programme was a result of pressure from the IMF. Certainly, it was perceived as somewhat demeaning for the GOB to have to engage an outside body to manage what had tradition-ally been the mainstay of its economy. The appointment was also conten-tious because of Booker Tate's relationship with Tate and Lyle who are effectively the market for all Barbados' sugar exports. These points aside, however, there may have been a degree of political expediency in the appointment as it allowed the GOB to distance itself from what is in some ways a 'no-win' situation which might well necessitate a number of politi-cally unpopular policy measures. In practice, the role of Booker Tate was somewhat disguised by the formation of a Barbadian management company, the BAMC, which would, in theory at least, have strategic control over the restructuring programme. In practice, however, both the day to day management and strategic planning have been placed largely in the hands of Booker Tate.

Booker Tate identified their primary objective as securing Barbados' pre-ferential markets, in particular the EU quota, but also delinquent quotas to the US. Accordingly, their plans were formulated around the need to pro-duce around 75,000 tonnes of sugar per year (ts/y). In the short term, some form of debt workout scheme was needed to allow the industry to continue to function. In the longer term, however, it was clear that to use Booker Tate's words 'quite radical' measures were needed if the industry was to remain viable (Booker Tate, 1993, Executive Summary). In a report produced in 1993, Booker Tate suggested two alternative strategies for restructuring the Barbados sugar industry. The first of these involved foreclosing on HIPs. This was not adopted, in part at least, because it was unclear whether the loan agreements between the BNB and the planters allowed for foreclosure. In any event, as Booker Tate acknowledged, action of this type would, in all probability, have resulted in extensive and long lasting litigation. The second strategy, which was subsequently adopted by the government, involved leasing the HIP's arable land for a period of 12 years.

Thus when the restructuring programme was commenced in 1993, BSIL, the HIPs and the government plantations all came under the direct, if temporary, control of BAMC and Booker Tate. BSIL was effectively wound up and the HIPs were leased for twelve years in line with Booker Tate's recommendations. Under the leasing arrangements, rents from the estates are paid directly to the BNB to service debts, but freehold title to the property remains with the individual plantation owners and they retain the opportunity to regain control of the land when the leases lapse in 2005, provided they are then in a position to repay any remaining indebtedness.

The substance of Booker Tate's restructuring plan centred around attempts to achieve efficiency gains. In particular, they argued that the plantations had been highly inefficient in their use of both labour and machinery and that considerable cost savings were achievable if practices were rationalised.

One of Booker Tate's first steps was to consolidate the HIPs and the government owned plantations into three operational units. This they suggested would allow them to use both machinery and labour more efficiently. Given the obvious inefficiencies existing on many plantations prior to Booker gaining control, significant cost savings may well be possible. Whether these will, in themselves, prove to be sufficient to allow the industry to become viable in the future is far from clear. As Booker Tate have recognised themselves, the problems are not simply technical. There are other, more 'intangible', factors underlying the collapse of the industry:

Perhaps the largest obstacle to a successful restructuring is intangible – the existing industry culture and its ability to change. The culture within the sugar industry is typical of a shrinking and unprofitable industry. Symptoms include:

a) The average age of employees is high;

b) Training and development of staff and succession planning are virtually non-existent;

c) Staff morale, team spirit and employee productivity are declining;

d) Investment in replacement equipment and new technology is low;

e) Effort is focused more on obtaining subsidies from government than improving operating practices;

f) Conflict between parties involved in the industry is frequent and sometimes prolonged; and

g) A poor image in the eyes of the public;

h) Most fundamentally, those involved in the industry are resistant to change. If the sugar industry is to continue in the longer term a fundamental change in culture is required . . . currently the industry is fragmented among many interrelated groups. Conflicts of interest exist between these groups which sometimes operate to the detriment of the industry as a whole.

(Booker Tate, 1993, vol.1. pp. 12–14)

Explanations of crisis in the sugar industry

As Booker Tate suggest, there are a range of explanations for the current crisis in the Barbados sugar industry. Those most commonly cited by a range of respondents within Barbados during 1994 are outlined below.

Currency fluctuations in early 1980s   When asked what caused the current crisis in the Barbadian sugar industry, almost all interviewees claimed that the fall of European currencies relative to the US dollar in 1981 played a significant role. The Barbados dollar is linked directly to the US dollar, but EU sugar payments are made in ECUs. When European currencies, and hence the ECU, fell sharply against the US$ in the early 1980s, sugar receipts in Barbados fell by almost 40 per cent within a few days. Although exchange rates returned to something nearer 1980 levels over a period of three or four years, the effects of the 1981 price shock are widely held to have been much longer lasting. One planter outlined the situation in these terms:

> In 1980 there were five Bajan dollars to one American dollar and it was the highest exchange value against the pound sterling and the Bajan dollar and the American dollar. When you sold your cane to England you were getting approximately B$1,750 per ton. By 1982 when the pound sterling had fallen so much that it was worth only three and at one stage two Bajan dollars we were getting around B$700 for a ton of sugar. That destroyed the industry. It was the exchange rate and nothing else at all. If you think that it is something else you are fooling yourself. Everybody will tell you all sorts of different things, but that was really what the problem was.
>
> (Personal communication)

The rapid and severe drop in incomes experienced by both the planters and BSIL in 1981 certainly did have a profound effect on the sugar industry in Barbados. That said, it is far from clear whether this factor can be seen as the sole or even the principal cause of the crisis. It may well be that a sounder, more efficient, industry could have weathered this price shock somewhat better than the Barbadian industry did. From this perspective, the events of 1981 served to expose underlying inefficiencies which, it might be argued, were more significant. Equally, however, the events in 1981 perhaps were crucial in that they persuaded many planters that there was no future for the sugar industry in Barbados. Once this had happened it became rational for planters to adopt strategies which recognised this. In practice, it may well have been these strategies which turned probability into inevitability.

Misappropriation of price stabilisation funds　The planters argue that moneys paid into a sugar industry price stabilisation fund were misappropriated by the government to finance the development of social infrastructure. The suggestion being that these funds (estimates vary as to what their value would have been by the early 1980s, but certainly over B$100m) would have gone a long way towards offsetting the consequences of the effective price collapse experienced in 1981. The government claims that for the most part these funds were used to develop infrastructure functionally related to the sugar industry, e.g. the new harbour which was built to handle sugar exports in Bridgetown.

Attitudes to the sugar industry　The long association of sugar with slavery in the Caribbean created a legacy of antipathy to the industry amongst black populations, which is widely held to still be significant. Certainly attitudes to agriculture and to sugar cane production in particular are far from universally positive in Barbados today. This cultural aversion is frequently cited as underpinning a number of problems including labour shortages and the malicious setting of cane fires. In practice, however, while considerable antipathy certainly does exist, the extent to which this is simply a legacy of slavery is perhaps debatable. On the one hand, plantation work is often hard and unpleasant and would in all probability always prove to be unpopular. Equally, it may be that pre-existing racial tensions effectively mask class structures which are, or at least are perceived to be, congruent with these.

Labour costs/problems　Almost all the planters interviewed claimed that rising labour costs have come to represent a major problem undermining the profitability and hence the sustainability of the industry. Labour costs do account for the majority of expenditure on plantations and sugar industry wages in Barbados are now significantly higher than those in many of its competitors' industries (see table 6.1). Wages have tended to rise in line with those in the tourism sector, and in response to a strong union. A high percentage of both agricultural and factory workers are members of the Barbados Workers Union. Barbados is a small island where the work force is easily unionised, and the size of the island and the nature of sugar production are such that it is relatively easy to promote effective industrial action. Moreover, the relationship between the planters and the union has often been antagonistic and strikes have been common. It is also the case that overall wage costs have been higher than necessary because the vast majority of plantations maintained work forces well above those needed to produce sugar efficiently.

Table 6.1 Agricultural wages in the Caribbean, 1992

Average daily earnings (B$)	Barbados	Jamaica	Guyana
Cane cutting	52	9	8
Field labourer	35	6	3
Skilled factory worker	72	13	5

Source: Booker Tate (1992, vol. 1:8)

Inability or unwillingness to modernise efficiently Given the relatively high labour costs and problems experienced in Barbados it is perhaps surprising that more effort has not been put into adopting modern, labour extensive production techniques. All plantations on Barbados have adopted technology to some extent. All, for example, use mechanised cane loading equipment and some form of mechanised cane transport equipment. Between 40 per cent and 50 per cent of the harvesting is mechanised (Booker Tate, 1993, vol. 2). However, a great deal of the technology which is present on the island is highly under-utilised – a feature of which Booker Tate were highly critical. Cane harvesters are extremely expensive pieces of equipment. Normally imported from Australia, such cutters cost something over US$250,000 in 1993. A number of plantations possess such harvesters which are only effectively used for two or three weeks during a five month harvest period. There has been little use of contractors or co-operation between plantations in the use of such equipment (Booker Tate, 1993, vol. 1). Some effort has also been made to develop a cheaper, smaller scale, single row harvester which would have been more appropriate to the scale of Barbados, but this initiative met with little success and has been abandoned. In practice, several barriers exist to effective and efficient mechanisation. Modern sugar production technology is very expensive and represents a considerable investment which normally requires external financing. The topography of many traditional cane producing areas is such that it is difficult or impossible to mechanise sugar production effectively, particularly in respect of harvesting. One widely expressed, and probably partially valid, opinion as to why the industry has not adopted modern production techniques is that many plantation owners are 'conservative' by nature and resist change.

Unwillingness to diversify Given the problems experienced in the sugar industry and Barbados' heavy dependence on food imports, a strategy of agricultural diversification, particularly the development of import substitution agriculture, would appear to be opportune in several respects. In fact, various official reports into the island's agricultural sector and successive National Development Plans have advocated such a strategy. Diversification would also appear to represent a sensible strategy for individual farmers. Agronomically it would alleviate some of the problems inherent in monocultural sugar cane production such as the build-up of pests and diseases. Economically, it would provide a better cash flow and reduce the vulnerability engendered by dependence on a single source of income.

However, whilst there has apparently been widespread agreement for some considerable time that diversification would be advantageous, relatively little progress has been made in this respect. Non-sugar agriculture has now become more important than sugar in terms of income generation, but almost 80 per cent of the agricultural land remains in sugar and large amounts of food are still imported. Most sugar producers on Barbados do grow rotational crops within the sugar cycle, but for the most part these are the root crops – yams, sweet potatoes, etc. – which have traditionally been produced and for which there is now only a very weak market. Relatively little progress has been made in introducing new crops. Most planters suggest that this reflects agronomic constraints. There may be some limited validity in this, but it would seem that a range of different crops could be produced. There are, however, other constraints. The domestic market for agricultural products is small, and much of this is accounted for by the tourist industry which requires not only particularly high quality but also security of production both of which might well be difficult to obtain on the island. The small scale of the island also makes for particular difficulties in developing a processing sector which would produce greater value adding and higher returns to the producers. These problems aside, it is the case that many landowners have interests in the tourist and food importation industries and thus have little incentive to undermine their profitable interests in food importation.

Absentee owners Absentee owners have long been a problem in Barbados, and plantations without resident owners have typically under-performed those which have been owner-managed (Watts, 1987:352). In theory, attorneys should be in a position to achieve efficiency gains through integrating the operation of several properties. In practice however, several interviewees questioned the effectiveness of this system. Attorneys were said to put their own properties first and to exploit other plantations in their charge, for example by overcharging them for the use of their own machinery. Perhaps, more significant than this, however, is the highly influential role played by a very small number of attorneys – just seven attorneys controlled some 63 per cent of the arable acreage on plantations in 1988. Thus, this very small group have had the potential to exercise a considerable degree of control on the industry.

Cost of inputs Almost all planters complain about the high cost and the low quality of inputs. Pesticides and fertilisers are said to be expensive and of dubious and unreliable quality. Particular concern is expressed over the cost of machinery and especially spares – which do appear to be inappropriately expensive. The government, however, suggest that most of the suppliers are closely linked with the plantation owners, either directly in that the owner-ship is the same, or through various informal networks. The corollary of this being that the planters are quite happy to make their money from these business interests rather than through sugar production *per se*. Not least because this arrangement facilitates a form of transfer pricing which allows the true profitability or otherwise of the plantations to be disguised and which thus can be used to substantiate the case for further government sup-port for the industry.

Agronomic problems Largely through a process of trial and error, the early planters on Barbados gradually developed a model of sugar cane agriculture, including for example the use of cane holes, which suited the particular environmental conditions found on the island. This traditional system was self-evidently sustainable in the sense that it allowed uninterrupted cane production for two or three hundred years. In recent years, however, a number of agronomic problems have emerged. Although monocultural sugar cane pro-duction is particularly susceptible to a range of diseases, Barbados has its own cane breeding station, and most people within the industry there are reason-ably satisfied with their work. Certainly new varieties have been developed to overcome diseases, such as smut, which have occurred in recent years.

Soil compaction is a major and widespread problem. In large part this is a consequence of mechanising various aspects of the production process. Even where cutting is still done by hand, fields are ploughed using tractors and cut cane is mechanically loaded onto some form of mechanised transport after cutting. Most observers suggest that although some degree of compaction may well be unavoidable, the situation is far worse than necessary because of a range of inappropriate practices in the use of machinery on many estates.

Soil erosion is a major problem in areas where sugar cane is no longer produced. The generally very thin soils of the island can be almost totally eroded during a single rainfall event if no adequate vegetation cover exists. Even where cane has been replaced with grass, slippage and other forms of soil movement are common. In the Scotland District, in particular, there have been severe problems of soil loss. Given the extreme vulnerability of Barbadian soils to erosion and the undulating topography of much of the island, it is perhaps somewhat curious that contour ploughing is not commonly practised in Barbados. Even on relatively steep slopes, upslope ploughing is the norm.

During the 1970s, controlled cane burning was introduced into Barbados in an attempt to address problems arising from the unregulated use of this practice. Although cane burning, which makes both mechanised and hand cutting easier, has been used effectively in other locations such as Queens-land, it proved to be highly problematic in Barbados. Both yields and sugar content fell dramatically during the period when controlled burning was used, and the practice was soon abandoned in favour of a return to green cane. The major problem with cane burning was probably that the trash, or trimmed leaves from the cut cane, was not left on the ground. A trash blanket serves to prevent evaporation and maintain soil moisture content, and when it is subsequently incorporated it raises the organic content of the soil. Despite the policy of not burning cane, cane fires are still often set. Workers being paid on piecework rate perceive an obvious advantage in being able to cut burnt cane quicker than green and this may well account for some fires. On other occasions the motivation for setting fires appears to reflect the tensions which exist between planters and cutters rather than any direct gain by the workers.

Extension services have been available to planters and farmers from a variety of sources in Barbados, including the Ministry of Agriculture, BSIL and the island's sugar cane research station. Relations between the Ministry's extension workers and many planters have apparently not always been particularly propitious. This probably reflects the different agendas of the Ministry and the planters, with the former attempting to promote a more diversified pattern of agriculture and the planters, for whatever reasons, not wishing to do this. Many extension workers argue that a large number of planters use inputs – fertilisers, pesticides and herbicides, etc. – very inefficiently. Knowledge and understanding of these products is said to be low. For example it is suggested that planters will commonly mix different proprietary brands and indeed different types of pesticide together. There is also a widespread suggestion that many planters have a poor understanding of how their machinery should be used. There is some evidence that inappropriate farming practices have contributed to quite widespread problems such as soil compaction.

Cost-price squeeze As profitability became compromised in the early 1980s most planters do appear to have economised on inputs, a response which would explain the dramatic fall in yields which has occurred at this time. The most significant method of reducing short-term costs in sugar cane agriculture, however, does not simply involve reducing the amounts of inputs such as fertiliser and pesticides used. With sugar cane there is a particular temptation to minimise short-term costs by ratooning for longer than the optimal period as this avoids expensive cultivation and replanting (Blume, 1985; 75). An optimal ratoon length is usually three or four years, but it is possible to ratoon almost indefinitely. This saves the cost of cultivating but both the yields and sugar content of subsequent harvests will be lower and lower each year. Thus, planters can get something of a free ride for a number of years, but such a practice is not sustainable in the longer term, especially where extended ratoons are not properly fertilised (de Boer, 1994:4; Wickham *et al.*, 1990:6). Moreover, it soon becomes unsustainable in the sense that it is unprofitable as returns fall below the costs of harvesting. Most respondents were unwilling to admit to cutting back severely on inputs or extending ratoons themselves, but most suggested that these practices were widespread amongst other producers.

To some extent, reducing expenditure appears to have been a reasonably rational response to the situation experienced in the early 1980s. As Booker Tate's manager pointed out, those planters who did try to maintain or indeed increase their output in order to offset falling prices rapidly found themselves in serious financial problems:

> Many of the people who went bankrupt first are those that tried desperately to maintain production, to replant their fields and to keep the industry going. They were the first to go under, to get bankrupt.
>
> (Personal communication)

Tragedy of the commons According to Booker Tate, the development of the crisis reflected a divergence between the individual and collective interests of the planters and the sugar industry as a whole:

> As an individual farm you could stay alive and run everything down without getting into serious debt, but if many of you do that you destroy the whole industry and that in a way is what happened. The production fell and the costs of the remaining output increased . . . it was a vicious circle.
>
> (Personal communication, Manager, Booker Tate)

Thus in something akin to a 'tragedy of the commons' situation, individually rational responses to problems proved to be collectively damaging. That said, whilst individual landowners certainly did adopt strategies defined by their own interests, and while these did contribute to the progressive crisis within the industry, it may well be that these strategies reflected individual interests which extended beyond the agricultural sector *per se*.

Loss of motivation by planters A large number of interviewees suggested that the problems of the sugar industry in Barbados have been compounded by a widespread belief that the industry has no real future. Thus planters are not motivated to invest in their properties or even to make any real effort to farm efficiently. The problem here is not simply that it is difficult to remain viable in the short term. The more perceptive planters, at least, are all too aware that the future of the ACP agreement is highly uncertain and that this is likely to have very profound consequence for the island's sugar industry. In practice, what seems to have happened is that many planters have adopted strategies which go some way beyond the passive response of not investing in their sugar interests and have apparently been positively removing as much capital from the plantations as possible and reinvesting this elsewhere.

124

Development gain, ineffective planning controls   Throughout much of the post independence period, considerable amounts of money have been made from development gain as, often high grade, agricultural land has been used for housing and tourist related developments. Apparently, this has dissuaded many planters from investing in sugar or being particularly concerned with developing their sugar interests as they intend to change the use of their land at the earliest opportunity. There has been quite considerable speculation in land related to the possibility of development gain.

In theory, the GOB has quite comprehensive development control powers. In practice, these have proved to be largely ineffectual (Carnegie, 1996). Most commonly, planning permission refused at normal levels has been granted under ministerial review. Accusations of corruption are widespread, but what tends to occur is perhaps something less than corruption *per se*. The situation is confused by the fact that a large number of plantations are owned by individuals or companies with interests in the tourism, construction and retailing industries. Patterns of individual and corporate ownership are complex and often involve senior political figures as well as landed interests.

To some extent the GOB appears to implicitly defend development gain as a legitimate objective through its claims that the current high level of debt owed to the BNB can be effectively expunged through the sale of further land for development. According to one eminent Barbadian politician:

> The Barbadian government has the capacity to recover that $200m at any time it wants to by the simple stroke of the pen allowing sub-division of some of the land that is now in sugar. And it moves the value of land from $2,000 or $3,000 to $10,000, so divide this by $200m and you will understand. There is a golf project, 400 acres of Westmoreland Plantation in St James, that has permission to develop. They claim that the receipts from this are going to be $600m. And Barbados has no capital gains tax and we don't have a development tax in the sense that you have in other countries.
>
> (Personal communication, former Prime Minister of Barbados)

Movement of capital out of sugar   As profitability in the sugar industry became compromised and new investment opportunities have emerged during the 1970s and 1980s, there appears to have been considerable movement of capital out of sugar into these other sectors. Accordingly, there has been under-investment in the sugar industry. Moreover, the government claim that much of the support afforded the sugar sector has passed straight through the industry to be reinvested elsewhere. There is considerable circumstantial evidence to support this contention. It is also widely suggested that money is not only invested in other sectors of the Barbadian economy but that much is exported to 'safer' locations such as the US and Europe. A former chief agricultural director in the Caribbean Development Bank summarised the situation in these terms:

> There were many years in which sugar would show a modest profit, sometimes no profit at all, but the subsidiaries were making considerable profits. Take for example the question of inputs into the sugar industry. The sugar industry should have been getting its own inputs – tractors and fertilisers, etc. – but instead of that companies were set up: Plantations Ltd, for example, the very name tells you. This was a company originally set up to deal in inputs for sugar. There were other organisations to deal with the storage and shipment of sugar after it was manufactured in the factory. All of this, in my view, should have been part of the industry. But these were siphoned off, but the same persons who ran the sugar industry actually ran these places as well, but they weren't part of sugar . . . the money that was being made out of sugar in those days was not being made out of crystals of sugar. All of these things were part of it and that was part of the problem.
>
> (Personal communication)

When asked if some people in the sugar industry were 'purposeful and objective in taking money out of the industry' one senior member of the Barbadian government suggested that:

> That is my impression. They used the system . . . in terms of business you could not call it illegal. If I can get money on soft terms and repay part of it and have the use of the rest of it, you see a lot of them used the moneys, they invested in other business activities and in some instances it was said they educated their children at the expense of the sugar industry.
>
> (Personal communication, Barbados Minister of Housing, Land and Environment)

The mechanisms through which money was allegedly transferred out of the plantations are well known. Receipts for sugar cannot be disguised, but income from non-sugar crops is almost exclusively in cash. Thus all inputs to the plantations whether these were for sugar or non-sugar crops were charged to the plantation, but cash income from non-sugar crops was habitually not accounted for. As one official suggested:

> My reading of the situation is this: after they saw the problems and realised that things were maybe not working according to plan, what has been happening, and I'm very adamant about this, they were planting other crops and filtering . . . this is very important. Under

126

the Sugar Act, the revenue from non-sugar crops was never taken, but the expenses were and the expenses were passed on against the background of sugar. When you put in a new crop, and you inter-plant onions or those other things, the manure was for the canes and not for those other things. It was classed as an expenditure for sugar, but when the crops were reaped the revenue did not go where it should, it went elsewhere . . . the labour cost on non-sugar crops is not accounted for, the labour cost goes to sugar, it goes against the expenditure for sugar while the profit goes straight into the pocket.

(Personal communication, senior official within the Barbados Ministry of Agriculture)

Certainly, it is the case that many of the planters also have interests in functionally linked commercial interests, which created the potential for a system of transfer pricing which would have allowed the real profitability of the plantations to be disguised. Moreover, this kind of activity would have been facilitated by the limited compass of the regulatory system on the island. Barbadian law does not require privately owned businesses with a turnover of less than B$1m to prepare full accounts and as most plantations fall within this category, hard evidence of the real accounts of these proper-ties is almost certainly impossible to obtain.

Poor government Given the nature of sugar production, there is an obvious need for some degree of regulation if only to co-ordinate the agricultural and industrial sectors of the industry. Various other factors including both environmental and economic risk and uncertainly, a number of potentially significant externalities and the high level of support afforded the industry also militate strongly for regulation. Many of the planters are openly critical of the government in Barbados. In practice, however, much of the regulatory framework which influences the Barbadian sugar industry is effectively external to the island, consider for example the ACP agreement and any changes which the EU may wish to bring about. Equally, it is clear that the Barbadian sugar industry is affected by elements of the mode of social regula-tion existing in Barbados which are not directly related to this sector. Con-sider, for example, the effects which the development of even a rudimentary welfare state on the island have had with respect to labour recruitment in the agricultural sector. Thus whilst the GOB remains the locus of most of the re-gulation which is formulated around the sugar sector, its potential and scope are necessarily highly selective, constrained and far from comprehensive.

Conspiracy theories In the first and last instance, most if not all plantations in Barbados which have become dysfunctional have done so as a direct result of high levels of debt. Whatever other problems the plantations may have had, it has been their inability to operate at a profit and thus to service their debts, which had made them unsustainable. Again, however, the situation may not be quite so straightforward as it at first might appear. Many members of the planter community argue strongly that the whole system of support provided for the sugar industry during the 1980s constituted a strategy by the GOB to encourage planters to over-extend themselves and thus to allow the state to appropriate the land. Certainly loans were made well beyond any reasonable banking criteria and there was never any realistic chance of them being serviced or repaid. And in practice at least half of the plantations did find themselves grossly over-extended with no hope of servicing their debts to the BNB. As one planter and attorney put it:

> They want to take over the whole thing. But why? I don't think they can manage my estate better than I can. And what am I going to do – sit on my arse? They are screwing me now . . . I think that in another couple of years it will be just like Rhodesia. They'll have chased every white farmer out. And then when the thing goes to hell, it will be the same as with Mugabe or whatever his name is, they'll have to beg them to come back. It will happen.
>
> (Personal communication)

However, the conditions under which the loans were granted did not allow the government to foreclose on the mortgages in the normal sense. Accordingly, although many plantations have been unable, or unwilling, to service their very high levels of debt, their estates have only been leased by the GOB, and the planters retain the option to regain control of the properties at the end of this period. One planter considered that this situation simply reflected the incompetence of the bank:

> If you are so foolish to do what the bank [BNB] did, which was to take out mortgages which did not give them the right to force the sale of the land, you'll never get your money back. Under that sort of business practice you're damned foolish, and that's what the bank has done. The bank got a mortgage which did not authorise them to sell off the property and they can't get their money back. The bank ought to be bankrupt, under normal business practice they would be.
>
> (Personal communication)

In reality, however, the leasing arrangement appears to undermine the suggestion that government support for the industry during the 1980s was part of a strategy to disenfranchise the planter community. This theory is also fundamentally negated by the fact that much of the money loaned to the planters by the BNB was never invested in the sugar sector. Had it been used as it was intended, it is highly probable that far fewer plantations would have become as indebted as they did.

The government theory is based on a contention that the planter community has quite objectively extracted as much support from the government as possible and then effectively expropriated much of the money ostensibly provided to support the sugar industry and used it for their personal benefit. There is considerable circumstantial evidence to suggest at least a degree of truth in this type of suggestion.

Senior members of both major political parties are quite overt and unequivocal in suggesting that the planter community has objectively sought, gained and subsequently misappropriated government money. As one Minister in the Barbadian government suggested:

> They don't look as if they are bankrupt, and surely they are not. And as for those who appear to be bankrupt don't ever look at that either – that's only for show. Rather than showing it off you hide it. They're still in control of 80 per cent of the land.
>
> (Personal communication, Barbados Minister of Trade)

The level of borrowing and debt accrued by many plantations appears to be well in excess of any figure which might have been accounted for by operational expenditure. A HIP debt of B$2.5m (US$ 1m), accrued over a period of less than ten years, equates to B$250,000 per year. Such figures are particularly difficult to account for given the fact that Booker Tate found virtually no assets – machinery, etc. – on these properties.

With little investment having been made on machinery, etc., during the 1980s, the only major expenditure of the plantations was on wages, which normally account for about 60 per cent of total outgoings. But this level of expenditure does not equate with the amount of borrowings which occurred. A typical plantation, employing ten full-time labourers earning a maximum of B$10,000 per year would have a wages bill for these workers amounting to B$100,000 per year. If a further ten workers were employed for the three months of the harvest this would add another B$25,000. If we add to this national insurance payments, etc., it seems unlikely that the total yearly wage bill for a typical plantation could be more than $150,000. Given that virtually no major purchases were made during this time, a generous estimate of other operational expenditures would be B$100,000 per annum. This would give a total yearly expenditure of B$250,000. Over a ten year period this would amount to B$2.5m, or roughly the figure which many plantations owe the BNB. Interest on these debts was capped at a maximum of 8 per cent, and even if we allow for the effect this would have, it is difficult to see how all of the borrowings could have been used to finance legitimate agricultural expenditure on the plantations. Balancing debt and apparent outgoings requires that these estates had virtually no income during these ten or so years and this simply is not the case. Most estates, whilst prices for their sugar may have been reduced for a period in the early 1980s, continued to produce sugar and to get paid for it. Most estates also produced some non-sugar crops, the incomes from which were not so affected by the currency problems.

From sustainability to unsustainability

There are then many possible explanations for the decline of the Barbados sugar industry. It may well be that the majority of these explanations are, in some sense, reasonable and accurate. What seems clear, however, is that specific technical explanations based on physical, agronomic or even labour supply problems are not, in themselves, complete explanations for either the demise of the sugar industry or for the range of 'unsustainable' events which have accompanied this. A more complete and powerful form of explanation needs to incorporate an understanding of (a) the external context in which the Barbadian sugar industry has operated and (b) the ways in which the unsustainability of the island's social structures have underpinned not only the collapse of the sugar industry, but also many of the more material and morally unacceptable forms of unsustainability which have come about.

The Barbadian experience demonstrates that the demise of a particular industry can be closely associated with a range of environmentally degrading and socially unfortunate impacts. In the Barbados case, these have included: accelerated soil erosion; potentially negative effects on the island's hydrology which may well severely prejudice water supply security; increased unemployment; a reduction in much needed foreign exchange earnings; and the misappropriation of capital needed for the wider development of the island. On the one hand, the restructuring process associated with the collapse of an established industry is itself almost inevitably going to be traumatic as traditional livelihoods, conventional land uses and established communities all become redundant. It is always going to be likely that both people and the environment will tend to suffer during such events. Not least because, as the Barbados case shows, relict industries will not necessarily be replaced by new and more productive economic activities in the same locations. Over and above this, however, the Barbadian case also demonstrates that the process of decline which pre-empts the final collapse of a particular industry may well tend to produce a range of unsustainable outcomes, some of which extend beyond the industry itself: unsustainable outcomes which, in this case at least, appear to have been the more or less direct results of attempts to address particular consequences of decline within the sugar industry. More succinctly, these are outcomes which are the result of particular actors or groups attempting to safeguard their own positions by adopting strategies purposively designed to protect extant capitals and class structures, the incidental consequences of which tend to be a range of degrading and destructive events. *Understanding these events as outcomes in a way which incorporates the whole range of causal factors involved is crucial to the achievement of sustainable development.* Such an understanding requires an appreciation of the objects and structures which give rise to the tendencies involved and the processes and mechanisms which produce their realisation. Also particularly important here are the institutional and social *conditions which legitimate and empower the mechanisms* involved. If unsustainable outcomes are ever to be avoided, it is vitally important to understand how modes of social regulation 'activate' causal mechanisms and thus allow actual unsustainable outcomes to be realised.

Context

The problems of the sugar industry in Barbados have to be understood and interpreted in terms of the guaranteed markets and the system of preferential prices within which the industry operates. The main quota is for 54,000 tonnes of sugar into the EU and the price which Barbados receives for this sugar is effectively determined by the 'A quota' price for European beet sugar. This averaged 27 US cents per pound in the 1980s, whereas the world market price for sugar averaged about 10 US cents per pound throughout this period. Thus Barbados received a premium of around 150 per cent for its sugar exports during the 1980s. And, the problems created by the 1981 currency fluctuations aside, Barbados has not been exposed to the extreme price volatility which typifies the international sugar market. It has enjoyed guaranteed markets and extremely preferential prices for virtually all the sugar it could produce. Even within this context, however, the industry has proven to be increasingly unsustainable.

In fact, the effectively stable conditions provided by Barbados' position within the ACP agreement defines a potentially useful context within which the 'unsustainability' of the island's sugar industry can be understood. Within this, there are two basic sets of explanations for the crisis in the sugar industry, and indirectly, for the related environmental and social impacts on development in Barbados. The first set of explanations are based on the various forms of absolute and relative inefficiency existing within the industry. The second set are based around the continued authority of the post-colonial social formation and attempts to develop modes of regulation which sustain its position despite the glaring 'cracks' that appear in such systems.

Inefficiency as a cause of unsustainability

To some extent, the current crisis in the Barbadian sugar industry can be explained in terms of the comparative disadvantages faced by Barbados and by the technical inefficiencies which have typified an industry within which there has been considerable resistance to the most necessary of changes. Certainly, the 'modernisation' of the Barbadian sugar industry was never accomplished in a particularly effective manner, and current inefficiencies may well have served to prejudice the profitability of the industry. But such explanations provide only a partial explanation of the current crisis, especially given the very advantageous context in which the sugar has been marketed. Moreover, if the substance of the crisis is falling output, technical inefficiencies in the production system cannot be the cause *per se*. What was essentially the same system 25 years ago produced almost four times as much sugar as that now being achieved. That said, these inefficiencies may have served to make the industry relatively less profitable than alternative economic activities, and this may well have been significant.

A similar argument pertains to role problems of labour supply and costs. It is widely argued in Barbados that the problems of ensuring an adequate labour supply have constituted a major factor in the industry's recent crisis; in regulationist terms, a major failure of the mode of social regulation. Clearly conditions have changed and the more direct forms of control and coercion adopted in the past are no longer tenable. But it is far from clear whether problems of ensuring adequate labour supplies can fully explain the sugar industry's problems. It is, however, important to appreciate that little which happens in Barbados can be properly understood outside the context of the racial and class based tensions which pervade Barbadian society. One significant question here is why did recruitment remain difficult even during periods of high unemployment? Certainly it appears that this reflected something more than the unpleasantness of the work or the wages available. Indeed, many of the labour problems experienced stemmed from the extremely poor relations which existed between the planters and the workers and the union rather than from any specific and material grievances regarding either pay or working conditions. Moreover in practice, the problems experienced in obtaining labour supplies on the island did not prove to be insurmountable. Labour was obtained from abroad when necessary. And beyond this, had the plantations adopted more modern production systems the need for labour would, in any event, have been drastically reduced. The extent to which the unsustainability of the Barbadian sugar industry can be blamed on operational inefficiencies is therefore somewhat debatable. The need for more efficient production methods has been apparent for decades, but little progress has been made.

The 'merchant-planter elite'

The direct political power of the 'plantocracy' has been gradually eroded during the twentieth century as the franchise has been extended, and ultimately with independence and universal suffrage. However, the indirect power held by what Beckles (1990) terms the 'merchant-planter elite' apparently remains. As direct political control over the island slipped away from the planter class, this group was able to extend the economic basis of its power. For some time, several companies have had both large landholdings and interests in retailing, importing, and in a range of other sectors including tourism. One Minister in the Barbadian government outlined his appreciation of the role played by the 'merchant-planter elite' in these terms:

> They have been able to buy these hotels and other kinds of investment and are trying to hold their control over the country – over successive governments. They have been trying to hold that tight rein of control. While the government is looking for political enfranchisement for all, the economic enfranchisement remains with the former group to this day. How do you maintain that? That's the question you should ask, how do you maintain that grip? You have to dig and dig and dig till you get right down to the truth . . . it's a subtle thing.
> (Personal communication, Barbados Overseas Trade Commissioner)

Another member of the Barbadian government commented in these terms:

> There was a group known as the 'big six' at one time. They were into sugar mainly, and they were into commission agencies in Bridgetown, and then they got into retailing activities, but none of them in those days ever got into any manufacturing. It was strange. We were here, government, political parties of both sides, pushing the idea of more manufacturing activities but they never got into this. They got into importing motor cars, owning garages – a quick, fast turn around. And they had a stranglehold on the import of everything under the sun which was brought into Barbados, and that is what it was. Their interests extended both ways, they had all the economic power.
> (Personal communication, Barbados Minister of Housing,
> Land and Environment)

In practice, the true economic and political influence of the 'merchant-planter elite' is extremely difficult to quantify. Commercial interests and networks are manifold and convoluted and business relationships are often informal and arcane. However, whilst the nature of the relationship which exists between the merchant–planter elite and the government is not as clear cut as some commentators would suggest, successive Barbadian governments have been overtly and vehemently critical of the merchant-planter elite and the ways in which they have used their economic and political power to the disadvantage of the majority of the population on the island. The suggestion is that this small community has been powerful enough to effectively subvert the political process. In practice, whilst political rhetoric is invariably populist, the reality is that this elite does appear to have been extremely effective in promoting its own interests. As one government minister suggested:

> What I know is that there were stages at which government's actions were constrained . . . its options were perhaps limited, but government never sought to control the activities of the private sector . . . except in more recent times when the central bankers had to restrain commercial activities because of the slippage of foreign exchange . . . there has never been any confrontation . . . the government has never had an adversarial relationship with them except in more recent times through the structural adjustment programme . . . what I am in fact basically saying is that none of the ministers of agriculture, apart from me, have ever had a confrontation with the sugar industry as far as I am aware. I was the first one that shouted out and I don't think they liked it.
>
> (Personal communication)

When explanations of the sugar industry crisis in Barbados are extended beyond purely technical matters, they tend to be polarised around the positions adopted by the government and the plantocracy, both of which interpretations are formulated around racial tensions on the island. In practice, racial and class issues are often conflated, which is unhelpful because the situation is far more complex than some commentators suggest. There is a significant black middle class, especially in the professions. Some black politicians have extensive commercial and landed interests. A large area of land is owned by blacks and has been for some time, and a number of new commercial interests which have been acquiring land on Barbados are not controlled by the Barbadian planter class.

Although the origins of the present day elite are clearly located within the plantocracy, the successors of the plantocracy are now more properly seen as a more purely economic elite. An elite whose position appears to have remained intact despite a whole range of antithetical and potentially damaging developments. Particularly the apparent declining profitability of the sugar industry. Irrespective of the precise composition of the current elite, however, it is clear that the unsustainability of the plantation system in Barbados is in itself merely a reflection of the unsustainability of a system of social relations which is a relic of seventeenth and eighteenth century colonialism. And here lies the basis of a much wider and much more meaningful set of unsustainable practices and events on the island.

In present day Barbados, there remains an elite class whose position became so fundamentally and clearly insecure that its members could not help but to be aware of their vulnerability. This group has striven to maintain its position – its wealth and privilege – in whatever ways it understands to be possible. To achieve this it has sought to maintain the effectiveness and viability of its economic base. And it has done this with a consummate disregard for the social and environmental consequences of its actions. Understanding how and why an increasingly incongruous and dysfunctional pattern of social relations has persisted for so long is centrally important to understanding the recent history of Barbados, and it is equally important to understanding the causality of a whole range of unsustainable practices and events which have occurred on the island.

Sustaining privilege

Until very recently the post-Colombian history of Barbados has been one in which a small elite group has prospered through the exercise of power founded on the ownership of land and the production of sugar. The industry produced great wealth, it made planters and merchants rich and it came to dominate the Barbadian economy to the almost total exclusion of almost all other economic activity. The plantocracy, a small group of historically white land-owners, were able to assure the reproduction of the conditions necessary for the production of sugar and thus the maintenance of their own wealth and their status. At times problems did emerge, particularly in ensuring the supply of labour necessary for sugar cane agriculture. At other times volatility in the world sugar market threatened the viability of the sugar industry. Around the end of the nineteenth century, Barbados and other colonial cane sugar producers were particularly threatened by the development of beet sugar industries in Europe and elsewhere. But throughout all of this, the Barbadian sugar industry was sustained and the status and privilege of the plantocracy was sustained along with it.

The second half of the twentieth century, however, has witnessed many changes in Barbados. Freedom from colonial control produced new development objectives, and ostensibly a relocation of political power. Independence did not, however, effectively change Barbados' markets for sugar which became institutionalised within the ACP agreement. That said, whilst it has been suggested here that the demise of the island's sugar industry needs to be explained in terms of largely indigenous factors, the wider context cannot be totally ignored. Income from the sugar industry may have been secured by the ACP arrangements, but it is clear to all concerned that the ACP arrangements are unlikely to continue in their present form beyond the medium term. Within this, it is equally apparent that when the EU discontinues or even merely starts to decrease its support, the Barbadian sugar industry will then be unprofitable however it is run. From this perspective, the future of the Barbadian sugar industry has been, at best, uncertain for some time. Once this far from profound realisation had been made and the inevitability of the situation accepted, the rational response of those involved in the industry is to do whatever they can to protect their own interests.

A particular group of people in Barbados have been able to sustain their own position, the value of their assets, their status and their power through the exercise of economic power. However, in itself, the economic basis of their power would hardly seem to be sufficient to explain what has occurred. Certainly, it would seem that the self-interested strategies adopted by this group were necessarily legitimated and empowered by institutions and values embedded in the society as a whole despite the fact that both the general population and successive governments have been highly unsympathetic to this group and its objectives. The ways in which such strategies are substantiated and capacitated within a mode of social regulation is of paramount importance to understanding of how many unsustainable events have come about and how they might have been avoided.

Historically the basis of the plantocracy's power was located unconditionally within the sugar industry, and this group strove to determine power structures and a political agenda which supported the sugar industry and hence the basis of their own wealth and power. Latterly, the elite class has been obliged to fundamentally reappraise their situation. The sugar industry and agricultural land are no longer the most viable investment opportunities in Barbados. The basis of capital accumulation has changed and so has the basis of wealth, power and privilege. The Barbadian elite class has responded to these new conditions by adopting new strategies. It has devised new goals and it has sought to influence the political agenda in new ways in order that it can pursue these goals more effectively. Or, perhaps more accurately, it has old goals that can now only be achieved in new ways.

The sugar industry has become unsustainable and many of the fixed assets associated with the industry will inevitably be devalued, but the basic class structures of Barbadian society have remained – they have been sustained. And, moreover, it would seem that they have been sustained through more or less purposive and objective strategies pursued by this elite group, including the marginalisation of sugar production. Whatever the populist rhetoric may be, the powerful in Barbados have given up on the sugar industry. New and more attractive investment opportunities have evolved and the Barbadian elite has responded to these developments. Their strategy has changed from constructing a political agenda which supported the sugar industry to one which allows them to maintain their wealth, status and power in new ways. Engagement with new accumulation processes, however, has necessitated the extraction of as much capital as possible from the sugar sector, including government support for the industry.

What seems to be so incongruous in the Barbadian experience is the degree of success which the elite class has had in averting the unsustainability of their own position. This is particularly surprising given the aspirations of post-independence governments to pursue a development path largely determined by the perceived injustices of the island's unfortunate history. Populist political rhetoric during the post-independence period has always been centrally concerned with undermining the position of traditionally privileged groups within Barbados – essentially the white plantocracy. Whilst it is easy to understand why such an agenda has formed part of all political manifestos in post-independence Barbados, it is equally clear that the actuality of government policy and practice has been largely determined by effective impotence and realpolitik than the singular pursuit of any developmental objectives. Successive post-independence governments could not ignore the ambitions of those who held and wielded economic power.

It would seem fair to say that the objectives of successive governments in Barbados have been genuinely progressive (Girvan, 1973; Pastor and Fletcher, 1991). The major political parties have always been broadly socialist and the immediate post-independence period was certainly perceived as being one of considerable opportunity and optimism. Almost thirty years after independence, there has been progress: living standards are relatively high and health and education provision are well developed by regional standards. These developments aside, however, the power and privilege enjoyed by the island's elite class remains as does its ability to exploit the island's resources and population. This group may be less obvious than the eighteenth century slave owners, or Beckles' merchant-planter elite, and the forms of exploitation and legitimation may now be more subtle and less transparent, but in essence the basic pattern of social relations established within the seventeenth and eighteenth century plantation economy remains.

The issue here is not simply that this group is able to exploit Barbadian resources largely for its own ends, or that the inequitable distribution of wealth on the island is in itself incompatible with notions of sustainable development. Also crucial are the unsustainable outcomes which have resulted from the processes through which the elite group has been able to sustain its own position and status. Here lies the basis of much that is unsustainable in present day Barbados. And what is important here, is not simply an understanding of the structures and mechanisms which have tended to produce such outcomes, equally significant is the social and political context which has allowed these tendencies to be realised in practice.

The incidentally unsustainable

In practice, the decline of the Barbadian sugar industry has been accompanied by a range of other 'unsustainable' practices and events such as accelerated soil erosion and negative impacts on the island's hydrological system. Many of these events are directly related to what has occurred in the sugar industry. In this sense, understanding the causality of the decline in the sugar industry informs our understanding of why these forms of unsustainability have occurred. What has transpired in Barbados is not simply that an increasingly unprofitable industry has no longer been able to maintain sufficient levels of investment to secure its future viability. Certainly planters have traditionally been more prone to engage in conspicuous consumption than to invest in new agricultural machinery, but what appears to have occurred has gone some considerable way beyond this. Large elements of the planter community have systematically and objectively transferred capital out of this sector. This has involved not just a lack of investment in new machinery and plant, but also recourse to definitively unsustainable practices such as extended ratooning. Extended ratooning is a profound and telling exemplar of what has been occurring on Barbados. The physical degradation of soils implicit in this practice (de Boer, 1981; Blume, 1985:75) mean that it is inherently unsustainable; first in that it can only proceed for a limited period, and second in that it precludes the future development of different forms of agriculture. It is so obviously an exercise in mining value, so patently a road to nowhere as far as agriculture is concerned, no rational planter who wished to remain in farming would engage in such a practice. Moreover, the quite widespread adoption of extended ratooning cannot be adequately explained by a general and overriding requirement for short-term cost savings – BNB loans to the plantation sector far exceeded the costs of normal cultivation practices. Extended ratooning makes very short-term super profit, but fundamentally undermines the potential for future agricultural production and future profits from agriculture. And in so much as ratooning can only be extended for a very limited number of years, such a practice is always likely to lead to other problems including the abandonment of formally productive agricultural land and associated processes of soil erosion. Thus both the environmental and economic sustainability of the land is severely prejudiced. Extended ratooning is the strategy of someone who has little intention of continuing in agriculture. In practice, however, extended ratooning was merely one example of the strategies adopted by the planter community during the 1980s and early 1990s. Indeed the approach adopted has predicated a range of materially and morally unsustainable outcomes.

Notwithstanding the specific environmentally and socially unsustainable outcomes associated with the demise of the Barbadian sugar industry, it is clear that the wider development potential of Barbados has also been prejudiced by what has occurred. Not only have there been unsustainable impacts on the island's physical environment, there has also been extensive, deliberate and apparently illicit misappropriation of the government funds allocated to the support of the sugar industry. Further to this, and perhaps most significant of all these events, is the fact that the new economic activities in which sugar industry capital is being invested may not be those which are most appropriate to the sustainable development of Barbados. A neo-liberal argument would suggest that these events have been brought about by the relative unprofitability of extant economic activities, and that investment in new, more profitable, forms of economic activity must be preferable. But it is not clear that this is the case here. For instance, it would seem to be manifestly reasonable to suggest that if sugar cane agriculture has to go, then the restructuring process should involve new forms of agriculture which (a) utilise the large areas of redundant agricultural land, and (b) would potentially be useful in that they would reduce Barbados' heavy dependence on food imports. A diversified non-sugar agriculture may be marginally less profitable than alternative economic activities, but that does not necessarily mean that it is unprofitable *per se*, or that it is not an appropriate and desirable development option for Barbados. Certainly the Barbados Ministry of Agriculture has argued for some time for a more diversified agricultural sector. However, whilst it now seems to be inevitable that sugar cane will soon no longer be grown on Barbados, the possibility that the vacuum left by the demise of sugar cane agriculture will be filled by new forms of agriculture seems to be highly unrealistic. The experience of other Caribbean sugar islands such as Antigua would suggest that this is a highly unlikely scenario (Government of Antigua, 1991). Areas of land in the Scotland District which have been taken out of sugar production in recent years have hardly been fully transferred to new forms of agriculture. Rather, they have been subject to abandonment which has often resulted in severe soil loss, a process which has only been partially offset by government soil conservation programmes.

One factor which militates against the adoption of new forms of agriculture is the fact that many landowners are quite content to see their land idle, as they content themselves with the prospect of future development gain associated with tourist and residential development. From their perspective the less productive the land is seen to be the better the case for development permission to be granted. Beyond this, it also seems that however rational and necessary import substitution agriculture may appear, especially given the indirect and non-economic advantages which it embodies, its development might well compromise the profitability of various well established food importation and distribution enterprises. Although many of these enterprises are effectively incorporated within an increasingly globalised food system, the majority are still owned by Barbadian or Caribbean companies. In practice, they are owned by individuals and groups almost all of whom are, or have in the past been, centrally involved in the island's sugar sector.

For the most part, the environmentally and morally unsustainable events which have recently occurred in Barbados have not been the direct or intentional result of particular actions. Nobody has deliberately encouraged soil erosion or promoted unemployment as a goal. Rather these developments have tended to be the indirect, incidental consequences of broader strategies concerned to address quite distinct matters. As the Barbadian elite group has responded to the changes which have increasingly come to threaten their position, they have acted to sustain their own interests. They appear to have been largely successful in achieving this. Unfortunately, this success has only occurred at the expense of the Barbadian environment and at considerable cost to the wider population.

Regulatory failure

Much that has happened in Barbados could be interpreted as regulatory failure. Specific examples of environmental degradation are easily identifiable, and many of the social consequences of current restructuring processes appear to be unsustainable in ways which transcend the purely moral constituencies of the term. Certainly apparently desirable non-economic aspects of development such as the amenity value of the island (though this may in fact have quite significant economic repercussions for the tourist industry) have been foregone. Indeed, more material forms of unsustainability have also occurred through processes such as soil mining. Moreover, it would appear that recent developments have already engendered new contradictions and forms of dysfunction which are likely to prejudice not only the sustainability of these developments themselves but also the wider development of Barbados as a whole.

Here it is useful analytically to differentiate between the more concrete instances of regulation – 'real regulation', legislation and the like – and what might be termed higher order elements of a mode of social regulation. Regulation, in the broad sense in which it has been used throughout this volume, is concerned with ensuring the conditions needed for a sustainable economy and a sustainable society. This involves something much more than merely addressing specific, concrete problems. The regulatory system in Barbados appears to have been inadequate in both the narrow and the broad senses in which the term is used.

To some extent, failure of regulation, in the narrow sense, may simply reflect the inadequacies of the regulatory system which one might expect to exist in a developing country. Specific measures have been enacted, for example, to control and redress soil erosion, to provide labour supplies for the sugar industry, and to develop a more diversified agriculture, but these initiatives have hardly resolved the problems being experienced. It may be that this failure simply reflects the impracticability of managing sustainable development. After the fact, *ad hoc*, end of pipe measures designed to address specific problems are never likely to be totally effective or capable of ensuring sustainable development. Indeed, the core of our argument here is that if sustainability objectives are to be achieved, regulatory systems need to encompass 'higher order' instances of regulation in ways which predispose development to modes in which unsustainable practices and events are unlikely to occur. However, it would be difficult to argue that what has occurred in Barbados over recent years has been particularly effective in this broad regulatory sense. Even where the objective of regulation is defined in the established sense in terms of the maintaining of the conditions necessary for social and economic sustainability, it is far from clear that even this has been achieved in Barbados. Whilst we must accept the regulationist position that modes of social regulation come about more through experimentation, struggle and conflict than through objective promotion *per se*, specific and apparently purposive actions in Barbados have clearly played an important part in the evolution of the mode of social regulation now existing there.

Although, in itself, the collapse of the sugar industry hardly constitutes an example of 'broad' regulatory failure, it could be argued more convincingly that regulatory failure was important in the collapse of the industry. Certainly the dysfunctional hostility which has occurred between the planters and the work force would seem to support such a contention. However, it is important to recognise that regulation, in this sense, is more concerned with *sustaining accumulation rather than any particular form of production*. Viewed in this way it is less clear that there has been regulatory failure. Much depends on whether the conditions needed for new forms of accumulation to function have been ensured, and again this is somewhat less than clearly the case. Within this, however, what is undeniable is that the Barbadian government invested large amounts of capital in vain efforts to sustain the sugar industry.

Tendency and realisation: structures, mechanisms and empowerment

A major theme emerging within this analysis centres around the recognition that the basis of much that is unsustainable in Barbados lies in the unsustainability of its relict social and economic formations and that this relational unsustainability has been translated into materially and morally significant outcomes. This process of translation whereby particular class structures are maintained at the expense of other aspects of development is crucial to understanding the causality of the unsustainable in Barbados.

The factors underpinning the unsustainability of extant social and economic formations on Barbados are multifarious. On the one hand, developments largely external to Barbados have changed the context in which economic activity on the island occurs. But the model of liberalisation and structural adjustment, promoted by the IMF and largely adopted by the GOB, has done little to alter the power of the local elite and has not succeeded in reorienting capital towards non-sugar agriculture. Moreover, whilst the ACP arrangements have provided some security for the sugar sector, the future of these is now uncertain. Technological developments have undermined Barbados' comparative advantage in sugar production and have led, for example through developments in air travel, to the expansion of other sectors such as tourism. Equally, however, it is also clear that the existing mode of social regulation has proved incapable of ensuring appropriate responses to these changes. The industry proved to be quite incapable of modernisation, not because there was no capital available to finance this, but because the landowning class either could not or would not fully adopt new practices or invest the necessary capital. Other forms of investment were clearly seen to be more lucrative and safer. However, the elite social group on the island has been able to sustain its own privileged position. It has achieved this through responding positively to changes in the economic and social environment from which it derived the basis of its status.

When an industry such as this ceases to function and traditional forms of accumulation become ineffective, the assets and resources upon which it has been based are clearly and unavoidably going to be devalued. And all things being equal, the capital employed in the industry and the status of those who own and control these resources is also likely to be devalued. But this is not what has occurred in Barbados. Whilst the processes involved have not yet quite run their full course, what has actually happened is that the assets and natural resources associated with the sugar industry are indeed being devalued, but the value of capital and the relative position of the capitalist class is being sustained. The largely insignificant is being sustained at the expense of the materially and morally consequential. Increasingly cognisant of the unsustainability of the sugar industry, the historical basis of their position, the owners of the industry have striven to extract every last cent from the ashes of its funeral pyre. Thus what has occurred is not simply that the resources involved have ceased to be useful in that they are no longer the basis of a productive industry. In practice, the resource base of the sugar industry has been systematically overexploited and degraded, with less and less concern being given to its reproduction. The rational pursuit of capitalist self interest has resulted directly in a range of unsustainable outcomes. Thus, the development dynamic has been central to the collapse of the Barbadian sugar industry and instrumental in promoting a range of unsustainable outcomes outside the industry itself. A direct relationship exists between the unsustainability of the particular social formation and a range of unsustainable events both within and outside the sugar sector.

The fact that the resources upon which the sugar industry had been based are no longer utilised, or more succinctly that they no longer have any economic value in sugar production, is in itself neither here nor there. Resources are dynamic, things become resources, things cease to be resources. The space and parameters for accumulation are thus constructed out of particular social formations and regulatory conditions. What is unquestionably significant from a sustainable development perspective, however, is that a number of resources, including formally productive agricultural land, have been degraded simply in order that the value of capital might be preserved. The dynamic redefinition of Barbadian resources has promoted, and has itself been reinforced by, the redirection and leakage of capital from the sugar industry. Some of the mechanisms involved here may or may not have been illicit, but the basic process involved is a totally legitimate exercise. Indeed not just in Barbados but much more generally, liberalisation of financial controls and terms of trade have been specifically designed to facilitate actions of this type. Thus we have an inconsonant situation where a mode of social regulation creates conditions which tend to produce unsustainable outcomes which subsequently have to be addressed through specific regulatory measures.

This interpretation of events in Barbados corresponds quite precisely with the realist explanation of unsustainable practices and events developed in chapters 2 and 3. As the viability of the extant socio-economic formation in Barbados has become increasingly prejudiced and unsustainable through time, strategies have been developed to defer this unsustainability, and as these have been selectively legitimated and empowered by the mode of social regulation a range of unsustainable outcomes has been realised. Figure 6.7 relates recent events in Barbados directly to this conceptual framework.

As the extant socio-economic formation on the island became increasingly insecure during the 1970s and 1980s, strategies were adopted to defer this unsustainability. Thus for example, *strategy 1* represents extended ratooning. This is effectively a mechanism for minimising production costs which allows increased profitability in the short term. But this practice results directly in soil nutrient depletion and damage to soil structure. In other words, it leads to the materially unsustainable outcomes indicated at moment *a* on the horizontal axis of the graph. *Strategy 2* is government subsidy of the sugar sector. *Strategy 3* is the transfer of capital out of the agricultural sector. Again both of these mechanisms are associated directly with materially or morally unsustainable outcomes.

From a neo-liberal perspective, government subsidy of the sugar sector is

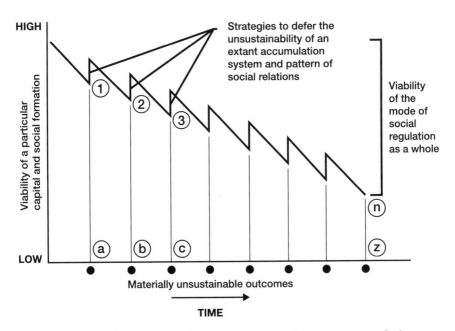

Figure 6.7 The mode of social regulation and unsustainable outcomes (Barbados)

inappropriate because in sustaining a sub-economic industry it is likely to produce a less than optimal pattern of development. There is some evidence that this has been the case in Barbados; a long history of subsidy and protection has resulted in manifestly inefficient production systems. These could be regarded as an unsustainable outcome in that overall welfare is clearly prejudiced by such a situation. The apparent misappropriation of government subsidies which appears to have occurred in Barbados must also have a material impact on the wider development of the island, and again could be construed as an unsustainable outcome. Certainly, it is unsustainable in the sense that it externalises the contradictions which have emerged within the sugar sector itself.

Similarly, the transference of capital out of the sugar sector appears to have produced a range of materially and morally unsustainable outcomes. Whilst this mechanism has served to sustain the economic basis of the landowners, it has simultaneously made a range of materially unsustainable outcomes related to the inappropriate farming practices almost inevitable. Beyond this, the effective transfer of government funds to individuals and private companies clearly prejudiced the wider development of Barbados which already had significant international debts and ongoing balance of payments difficulties.

In practice, the relationship between the structures which give rise to these mechanisms and the actual outcomes produced is governed by the nature of the mode of social regulation. Objects and structures such as those embodied in the global sugar economy, the ACP protocol and the pattern of social relations and the land tenure system existing in Barbados, tend to produce contradictions and tensions which undermine the profitability of sugar production and thus encourage the types of strategy adopted. But the mechanisms which these strategies embody are only 'activated' when they are legitimated and empowered by the institutions and values which constitute the mode of social regulation. Thus in this case, at least, there are structurally defined tendencies which predicate particular mechanisms. In turn, these tend to produce a particular type of outcome. Thus it would appear that an 'internal' or 'necessary' relationship exists between these objects and structures and the materially and morally unsustainable outcomes which tend to occur, and indeed in this case actually have occurred. However, this relationship within which 'relational unsustainability' is translated into 'material unsustainability' is itself fundamentally conditioned by the particular mode of social regulation. And a crucial point here is that whilst the mode of social regulation legitimates and empowers strategies which are temporarily 'successful', new contradictions and tensions will continue to emerge and new strategies will have to be put forward to address these. Because the contradictions generated are likely to become increasingly profound, *strategy n* is likely to involve more extreme forms of exploitation than earlier strategies. Thus, *event z* is likely to be a profoundly unsustainable outcome.

To understand the causality of these unsustainable outcomes from a realist perspective it is necessary to understand, first, what mechanisms are involved; second, the objects and structures which give rise to these; and third, the context which substantiates these processes. In practice, this context is defined by those elements of the mode of social regulation which legitimate and empower the mechanisms involved. Various specific mechanisms appear to have been involved here, but a key causal mechanism has been the transfer of capital out of the sugar sector, both directly and indirectly. Thus we might infer, or 'retroduce', the existence of structures and objects which are necessary for this to have occurred. There must have been, for example, a particular form of property rights for this to have been possible. In this case, landownership patterns and the private property rights constituted in the plantation system have allowed individuals and companies to pursue particular strategies. But the very fact that these strategies have been successful suggests that other structures constituted in the institutional and social conditions existing in Barbados have also been significant. Thus we could see some significance in the ineffectiveness of the planning control system. And, perhaps more importantly, in the ways in which the landowning class have been able to manipulate the political agenda to their own advantage.

In Barbados an extant socio-economic formation – the sugar sector and the pattern of social relations which historically has been associated with this – has become increasingly prejudiced and crisis prone. This has created tendencies to overexploit environmental and human resources. These tendencies have frequently been realised in practice because the mechanisms involved have been legitimated and empowered by the mode of social regulation. The mode of social regulation existing in Barbados prioritises and objectifies strategies designed to sustain the value of capital and through this to sustain extant class structures. The mode of social regulation, therefore, legitimates increasingly and ultimately unsustainable forms of exploitation.

As the Barbados case shows, the realisation of particular events reflects not just structurally defined tendencies and contingencies, but also *the conditions in which the mechanisms involved operate*. Faced with the unsustainability of their position the Barbadian elite class embarked, quite objectively, on courses of action which would protect their status, the more or less direct consequences of which were always likely to result in a range of unsustainable outcomes. But, crucially if these unsustainable outcomes were to be actually realised, the processes and mechanisms set in place needed to function effectively. Given the nature of these processes, this could only occur if they were legitimated and empowered by the prevailing mode of social regulation. Thus it follows that if these processes and mechanisms were not empowered, then the tendencies to the unsustainable which they involve would not have been realised.

Successive governments in Barbados have become increasingly acquiescent regarding neo-liberal approaches to development. In part, this may reflect the country's weak financial position and pressure from the IMF. Equally, however, policies such as the liberalisation of commercial activity, the formation of a Barbados Securities Exchange and the development of an offshore financial services sector are all increasingly seen as necessary for the economic survival of the island. Thus whilst the processes through which the elite has been able to sustain the validity of its own economic position has certainly reflected the inadequacy of the island's regulatory system, these processes have also been facilitated by the adoption of increasingly neo-liberal policies, including moves to liberalise financial controls in Barbados.

The current mode of social regulation on Barbados has served to legitimate and empower mechanisms which sustain extant class structures and the privilege of an elite community, but only at the cost of conditioning development to the unsustainable. The outcome is a whole range of unsustainable practices and events which cannot be effectively addressed in their specificity. For instance, environmental management or agricultural diversification schemes will, at best, modify this type of outcome, they will not prevent them. The achievement of sustainable development requires that development is reconditioned in ways which avoid the unsustainable rather than predispose it. In practice, this means that, amongst other things, those elements of the mode of social regulation which legitimate flows of capital through economies need to be re-evaluated.

This analysis supports the contention that unsustainable practices and events need to be, and can be, understood and addressed as the outcomes of economic and social processes and the institutional and social conditions in which these occur. The interpretation of recent occurrences in Barbados which has been developed in this chapter shows a more or less direct link between the unsustainability of the extant socio-economic formation on the island and a range of materially and morally unsustainable practices and events. In particular, it appears that the mode of social regulation in Barbados, which involves both traditional and newly emerging regulatory forms, has legitimated and enabled strategies which prioritise, objectify and prescribe flexibility to the value of capital and the reproduction of extant class structures. In this way, the mode of social regulation in Barbados has conditioned development to the unsustainable.

THE REGULATION AND RE-REGULATION OF THE AUSTRALIAN SUGAR INDUSTRY

Introduction

This chapter begins with a short history of the Australian sugar industry. The current structure of Australian sugar production is then outlined in some detail, with particular attention being paid to the nature of the sugar industry regulatory system which has been in place for most of the twentieth century and the deregulatory process currently being enacted. Case studies of two sugar producing regions in Queensland are used to illustrate the range of environmental, agronomic, economic, social and moral forms of un-sustainability which have been associated with the development of the industry.

The second half of the chapter interprets these events in terms of the approach outlined in chapters 2 and 3. The discussion focuses on the contradictions and tendencies to dysfunction which have emerged within this sector, and the strategies through which these have been addressed. Our analysis focuses on the ways in which the development of the industry and the impacts of this development have been conditioned within three key phases of regulation and re-regulation: a plantation phase subject to imperial regulation up to *c*.1900; a regime based on family farming subject to national productionist regulation which pertained from the turn of the century until the 1980s; and a newly emergent regime reflecting the deregulation and re-regulation of the industry over the last decade. In this context, we seek to demonstrate that the modes of social regulation associated with each regime have tended to condition development towards materially unsustainable outcomes.

The Australian sugar industry

Australia is atypical of most sugar cane growing countries in that it is a progressive, developed country with a high wage economy. Over 80 per cent of Australia's total population of 17.1 million live in the state capitals and other urban centres. Population densities are extremely low in most rural areas. With a total land area of 7,687 square kilometres Australia is a very large country, but climate, especially rainfall patterns, severely reduce the amount of land suitable for most forms of agriculture.

Australia has been independent since 1901 and has a federal system of government with legislative powers vested in both Commonwealth and state parliaments. Although now considerably less important than they were in the past, primary industries remain central to the Australian economy. Primary commodities accounted for 53 per cent of the value of Australian exports in 1990 (World Bank, 1992). Within this, sugar accounted for approximately 6.5 per cent of the value of Australian farm based exports in 1991 (Australian Bureau of Agricultural Resource Economics [ABARE], 1991:4).

Commercial production of cane sugar in Australia did not begin until the 1860s. The first crop produced in Queensland was planted in 1862, but drought and financing problems meant that expansion was initially slow with only 338 tons of sugar being produced in 1867. By 1874, however, there were 71 operational mills producing over 4,200 tons of sugar (Graves, 1993:12). Although the Australian sugar industry had its origins in Northern New South Wales and Southern Queensland, these areas were climatically less than optimal for sugar cane agriculture and the industry soon expanded into warmer and wetter areas further north. The port of Mackay in Queensland, for example, rapidly became an important centre of sugar cane agriculture and milling with over 1,000 acres of cane being brought under cultivation and 17 mills being constructed between 1870 and 1874. By the last decades of the nineteenth century, sugar production had become firmly established in a number of agronomically conducive pockets along Australia's north-east coast.

Despite the fact that the establishment of the Australian sugar industry post-dated that in Barbados by over 200 years, a number of close parallels exist in the patterns of development. One of these was the plantation. As in Barbados, early sugar cane production and processing in Queensland was organised around a plantation system. Early producers in Queensland may have merely sought to emulate the models existing elsewhere in the world at that time. But, it is also the case that the nature of sugar production at that time predicated the type of systems which might be used. Few in nineteenth century Australia were in a position to establish a sugar production enterprise which required land, an expensive mill, and access to large amounts of labour. Thus, in practice, the technological and social conditions existing in the mid-nineteenth century may well have served to determine not only the type of production system used in the sugar industry, but also to define the pattern of social relations upon which it was based. However, whilst the plantation has for the most part remained the basis of production on Barbados, it had essentially ceased to exist in Australia by the end of the nineteenth century. From that time onwards sugar production in Australia has been based on large numbers of family farms.

The cardinal problem underlying the collapse of the plantation system in Queensland was that of ensuring adequate labour supplies. Early plantations in the Caribbean had used slave labour and, whilst slavery *per se* was never an option available to Australian planters, attempts to maintain an adequate labour force were to become increasingly coercive. Throughout most of the plantation period in Queensland (*c*.1850–*c*.1900), the industry relied heavily on the use of foreign labourers. Initially Chinese and other Asian labourers and, to a lesser extent, European workers were recruited, but the principal body of labour used during the second half of the nineteenth century consisted of Pacific islanders. These islanders were recruited by specialist traders on their home islands and contracted to work for specific periods at predetermined rates of pay. By the 1880s there were over 5,000 'kanakas' working on Queensland cane farms. Pacific islanders, coming as they did from radically different cultures to that in Queensland, did not constitute an ideal workforce, but they were at least sufficient to allow the industry to function and indeed expand. Having made a quite considerable investment to secure a workforce, planters were committed to ensuring a return on their investment. The conditions in which the Pacific islanders worked were typically extremely poor, with various forms of coercion, including violence and intimidation, becoming increasingly common as the century progressed (Graves, 1993). Within a relatively short period, however, recruitment and retention were to become increasingly difficult. The costs of recruiting these labourers rose progressively and it eventually became difficult to obtain new workers at any price. Moreover, the treatment of these workers produced considerable moral outrage in Australia. This led to legislation controlling their conditions of employment and, eventually, to a statutory ban on the importation of Pacific island labour into Queensland in 1904. Thus the trade in and use of 'kanaka' labour came to an end at the turn of the century with all but 1,600 islanders having been repatriated by the end of 1907 (Saunders, 1982; Shlomowitz, 1982; Graves, 1993). With this supply of labour no longer available, the plantations were unable to function and the transition to a new system of production became unavoidable. As Graves puts it:

> Here was an industry highly stressed by the need to optimise surplus extraction under extremely adverse circumstances, through the exploitation of a labour force which was not merely inexperienced but was largely unaccustomed to the rigorous time and work discipline demanded by capitalist agro-industry. These acute stresses not only prompted the introduction of 'progressive' work and organisational practices, but also an extremely labour-coercive system, which not only appealed to a range of legislative and social controls, but which relied on the pervasive threat of violence against plantation workers. How effective these strategies were, is of course, open to question for . . . Queensland's plantations were found wanting and

were rapidly replaced by a less objectionable, and more dynamic system of production. Powerful though the institution of the plantation was, it was insufficient to serve all the industry's needs.

(Graves, 1993:132)

Although the plantation system in Queensland was rendered unsustainable because of labour supply problems, other concurrent developments were also militating for change. Developments in milling technology, for example, were quickly rendering the established, plantation based, cane processing systems obsolete. As with Barbados and indeed in almost all sugar producing regions, the inherently superior efficiency of large centralised mills meant that a separation of the agricultural and industrial components of sugar production was inevitable (Deere, 1949). With land on redundant plantations now widely available the favourable conditions for the entry of many smaller cane farmers suddenly became established in Queensland around the turn of the century. The demise of the plantations led directly to a production system based on family farming allied to the use of centralised factories. This structure of production has essentially remained intact throughout the past 90 years.

While the basic geography and structure of the industry had already been defined by the turn of the century, the area under cane has increased more or less consistently since then as the amount of local land used for cane has periodically been expanded in established sugar producing regions. As this expansion progressed during the inter-war and immediate post-war periods, individual cane farms tended to grow in size and many were again faced with the problem of obtaining adequate supplies of labour, especially during the harvest season. And again, many came to rely on immigrant labour, in this instance usually of European origin (see for example, Kerr, 1988; Manning, 1983). One enduring legacy of the use of immigrant labour on Queensland cane farms during the 1940s and 1950s is the present day concentration of ethnic minority groups in several cane producing areas. In many cases immigrant workers have become established cane farmers in their own right. In the Bundaberg area, for example, there are significant numbers of ethnically Italian cane farmers and in the Mackay area there is a substantial minority of farmers of Maltese descent. Although by the early 1990s, most of these farmers are second generation Australians, ethnic groupings often remain clearly defined.

However, the period of reliance on off-farm labour which occurred during the inter-war and immediate post-war periods was short-lived as increasing mechanisation made sugar cane agriculture an increasingly labour extensive exercise in Australia (Manning, 1983). Rather than promoting a radically different structure of production, this transformation served to reinforce the position of the family farm as it allowed individual farmers to increase production without recourse to off-farm labour. The viability of the family farm was further supported by what was to become the highly intensive use of chemical pesticides and herbicides which also tended to make production less labour intensive. The present day Australian industry is probably the most technologically and chemically intensive and reliant sugar industry in the world: the epitome of capital intensive high input–high output agriculture.

Australia became a net exporter of sugar during the 1920s and by the post-Second World War period, the majority of sugar production was being exported. Traditionally, Britain had always been the primary market for Australia's sugar exports and in 1954 this trade was formalised under the terms of the (British) Commonwealth Sugar Agreement. This was an arrangement which clearly had a number of benefits for the Australian industry:

> Historically the industry has viewed the BCSA with a mixture of gratitude and sentiment. It certainly played an important role in the development of Australian sugar production during the fifties and provided a guaranteed outlet for about a quarter of Australia's exportable surplus in the sixties, when prices were low and over production made sales difficult. The BCSA also gave a price guarantee which up to 1973, could be said to have provided over the years, a better return than was possible for sales on the open market. From 1966 to 1971 the negotiated price was £43.50 per ton for the quota of 335,000 long tons, which was raised to £50.00 in 1971 and to £61.00 in 1974. The 1966–71 price gave an approximate return to the industry of A$110.00 and at 1974 prices a return of about A$99.00 per ton.
>
> (Lance Jones and Co., 1975:27)

When the Commonwealth Sugar Agreement expired in 1975, Australia was the only former party to the agreement not to be included in the ACP Protocol of the Lomé Convention. Thus from 1975 onwards Australia had to find markets for substantial quantities of sugar exports within the global sugar economy. A large proportion of subsequent exports took place under a series of bilateral arrangements with importing countries, most notably with Japan. However, whilst these agreements may have created some degree of price stability for the industry, they never included any great premium over prevailing world market prices (Queensland Sugar Corporation, 1992a).

By the end of the 1980s Australia was exporting the majority of its sugar to nine main destinations. Japan accounted for 20 per cent of all exports, Malaysia 19 per cent, Canada 15 per cent, South Korea 13 per cent, the USSR 11 per cent, China 7 per cent, Singapore 6 per cent, the USA 6 per cent and New Zealand 3 per cent (Sugar Board, 1991). Between 25 per cent and 30 per cent of these exports were covered by long-term contracts which existed with Malaysia, South Korea, China and the Soviet Union (ABARE,

1991:17). In practice, however, the situation is highly unstable and several developments outside Australia are likely to redefine its exports markets. The rapid growth of the Thai sugar industry has made Thailand an extremely important player within the regional context and further expansion seems likely. The role of Cuba is also highly significant and would become even more so if that country's relationship with the United States were to improve in the future. According to the Senate Committee on Industry, Science and Technology (SCIST) (1989:12), the Australian sugar industry has probably been more exposed to world prices than any other major producer. Ranking the exposure of exporting countries on a scale of zero (no exposure) to four (complete exposure), only Australia and Thailand rated a score of three.

The present day Australian sugar industry

By the late 1980s, Australia was producing well over 4 million tonnes of sugar per year, which amounted to around 3.5 per cent of total world sugar production. Approximately 80 per cent of total raw sugar production was being exported at this time. Australia is the world's third largest sugar exporter with around 10 per cent of world trade, after Cuba (24 per cent) and the EU (20 per cent) (F. O. Licht, 1993). Australian sugar exports had a value well in excess of A$1 billion in 1990 (Queensland Sugar Corporation, 1991:6). (At 1994 exchange rates, A$1=c.£0.50, c.US$0.75.)

Seldom occurring more than 50 kilometres inland, sugar cane agriculture occurs in a number of pockets along Australia's east coast (see figure 7.1). The southernmost cane growing areas lie in the north-east corner of New South Wales and the most northerly, 2,100 miles away around the town of Mossman at the foot of the Cape York Peninsula in Queensland. The total area of land devoted to sugar cane in 1990 was 403,000 hectares, with 95 per cent of this total occurring in Queensland (Sugar Industry Commission, 1992:23). Sugar is the second most valuable agricultural commodity produced in Queensland with only beef cattle being more significant (ABARE, 1985:4).

Rainfall and temperature both vary significantly between the different sugar producing regions. In the most northerly regions rainfall averages over 2,300 mm per year and average annual temperatures are around 27 °C. In New South Wales conditions are very marginal for sugar cane agriculture with mean annual rainfall only just over 1,000 mm and average temperatures of only 19 °C. In practice, rainfall patterns are the key environmental constraint faced by sugar cane agriculture in Australia. In some areas rainfall patterns are barely adequate to sustain production. Even where rainfall is more plentiful, it tends to be unreliable and drought is not uncommon. Irrigation is a widespread, though far from universal, feature of many cane growing areas in Southern and Central Queensland.

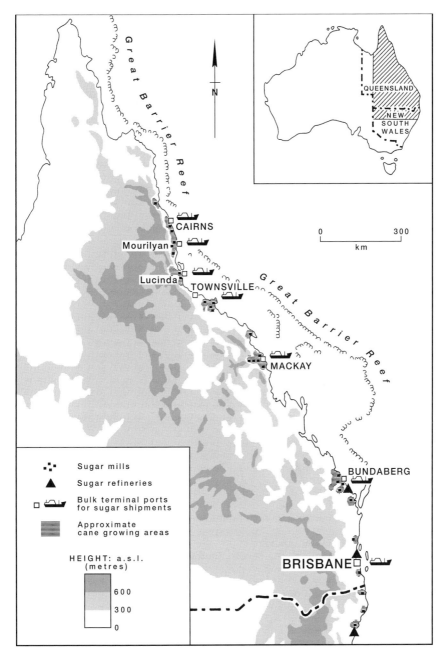

Figure 7.1 Australian sugar producing areas

In practice, the considerable geographical extent of the Australian sugar industry is useful in that it gives the industry a degree of resilience to a range of different forms of stress such as drought, disease or industrial action. As Powell and McGovern put it:

> In the international market, the Australian industry is known for its stable production of a quality product – a most desirable attribute for a product widely used as an input to a range of further manufacturing processes. Such a reputation does reflect a generally uniform use of appropriate technologies but it also rests on the strength arising from the diversity present. For example, rainfall varies amongst the regions and in a region from year to year – but total production remains relatively stable despite dry patches here and there. Also within the industry there are enterprises that are doing well and others that are going broke steadily or spectacularly. The heterogeneity of the industry and its sectors lies in its overall stability and performance.
>
> (Powell and McGovern, 1987:9)

In 1991, sugar cane was grown on just over 6,000 farms in north-eastern Australia. Most of these farms grew between 30 and 90 hectares of cane with the mean area of cane per farm being approximately 65 hectares (Sugar Industry Commission, 1992:23). In recent decades there has been a consistent increase in the size of production units. However, although the area planted in sugar cane increased from just over 300,000 hectares to 360,000 hectares between 1970 and 1986, the number of cane farmers fell by around 1,500 during the same period (Powell and McGovern, 1987:17). Most Australian cane farms produce nothing but cane. Traditionally, rotational crops are not grown and often the entire farm is used for cane. Few cane farms employ any non-family labour although the majority use contractors for much of the cultivation and harvesting work. Some 4,000 contractors are primarily engaged in sugar cane agriculture (Sugar Board, 1991:27).

Australia has been at the forefront of innovation of specialised technology for sugar cane agriculture for several decades. The first practical cane harvester was developed in Central Queensland and a Bundaberg based company remains the world's leading supplier of cane harvesters. Indeed, the use of modern technology has become deeply ingrained in the culture of Australian sugar cane agriculture. Virtually all the Australian sugar cane harvest is cut mechanically. The transition to mechanised harvesting occurred mainly during the 1960s; in 1964 24 per cent of the total crop was cut by machine, by 1973 99.6 per cent was being harvested mechanically (Lance Jones and Company, 1975). However, whilst all the crop is cut mechanically few farmers own their own harvesters. In 1990, there were some 1,300 mechanical cane harvesters in Queensland, one for every five farms (Queensland Sugar Corporation, 1992b).

A sophisticated cane transport and sugar handling infrastructure exists throughout the cane producing areas of Australia. In Queensland there are, for example, some 3,900 kilometres of specialised narrow gauge railway which is used to transport cut cane from the fields to the mills. A number of bulk sugar export terminals exist along the north-east coast of Australia. A total of 70,000 hectares of sugar cane land is irrigated in Queensland (Queensland Sugar Corporation, 1992a:7). Several areas benefit from specially constructed irrigation schemes. The Bundaberg–Isis irrigation scheme, for example, provides irrigation water for several hundred farms in the Bundaberg district who have no access to either the Burnett river or groundwater supplies (Hungerford, 1987). Australia also has a highly developed sugar industry research and development infrastructure spanning both the agricultural and milling sectors. In practice, cane farmers receive extension services from several agencies with well defined, but sometimes overlapping remits.

Most of the sugar mills which exist in Queensland today were established in the late eighteenth and early nineteenth centuries. Since then, progressive gains in mill productivity have consistently outstripped the growth in cane production, and consequently the number of mills has tended to decline. This trend continues today – five mills closed between 1986 and 1991. By the end of 1991 there were 28 raw sugar mills in Australia, 25 of which were in Queensland and three in New South Wales. The three mills in New South Wales were owned by a single grower co-operative. Of those in Queensland, seven were owned by CSR Ltd (formally Colonial Sugar Refiners Ltd), six by Bundaberg Sugar, and four by the Mackay Co-operative. Most of the remainder are also owned by local grower co-operatives. CSR not only owns a number of mills, it is also by far the largest sugar refiner and wholesaler in Australia, accounting for around 95 per cent of domestic sugar sales. This company also acts as agents for the Queensland government in the marketing of sugar overseas.

One official estimate suggests that in 1983 between 45,000 and 60,000 people were either directly or indirectly dependent on the sugar industry for full-time employment in Queensland (SCIST, 1989:6). In 1991, only 900 non-family workers were employed full time on cane farms, but the milling sector employed some 6,000 people during the harvest season (usually around 21 to 22 weeks during the second half of the year) and somewhat less than 5,000 during the remainder of the year (Sugar Board, 1991:27). Although the numbers directly employed in the Australian sugar industry are not particularly high, they are significant in the context of rural Australia. The total population of Queensland is approximately four million, and almost three quarters of this total live in Brisbane, the state capital. Moreover, the sugar industry is geographically concentrated into a number of small areas which have a very high degree of dependence on this one industry. As one recent government report pointed out:

> The sugar industry has strong regional effects on employment in some areas. A number of towns are essentially 'mill towns', for example Mossman, Hambledon and Tully. A number of regional areas are similarly dependent . . . the Burdekin community is largely dependent on sugar cane for its income with approximately 80 per cent of income in the district directly attributed to the sugar industry. Similar claims are made about other areas.
>
> (SCIST, 1989:6)

Whilst the sugar industry remains extremely significant at both regional and state levels, the total number of workers involved has been falling in recent decades. A trend to declining employment numbers became established in the late 1960s and has continued to date. The number of families involved in cane farming fell by around 20 per cent between 1970 and 1985, and the number of people employed in the milling sector fell by over 10 per cent during this same period (Powell and McGovern, 1987:29). Similar reductions in employment numbers in both the agricultural and milling sectors have persisted into the 1990s (Sugar Industry Commission, 1992:27). Although the Australian sugar industry now employs less people than it did in the past, production has increased steadily in recent decades. During the last 30 years output has more than doubled, rising from 1.3 million tonnes in 1960 to over 3.3 million tonnes in 1990, see figure 7.2. Much of the year on year variation in sugar output indicated in figure 7.2 results from climatic effects on production. 'Plant crops', that is the first crop produced after cultivation, are the most susceptible to drought and under adverse conditions can fail totally. Ratoon crops are more resilient, but the c.c.s. or sugar content of the cane is often low when rainfall patterns have been sub-optimal.

Most of the overall increase in output after 1960 occurred as a result of periodic increases in the area of land being used for sugar cane agriculture (see figure 7.3). Throughout this period, statutory controls meant that sugar cane could only be grown on land 'assigned' to that purpose by the Queensland government. But, in practice, a fundamentally productionist ethos led to periodic increases in the amount of assigned land. Total land assignments which stood at around 300,000 hectares in the early 1970s, rose to around 330,000 hectares in the late 1970s and again to around 360,000 hectares in the early 1980s.

During the 1960s and early 1970s, there was also a trend to increased yields. However, as figure 7.4 shows, further increases in yields proved to be difficult to achieve after around 1975. This had to be the case because the potential to further increases in yields through technological advancement

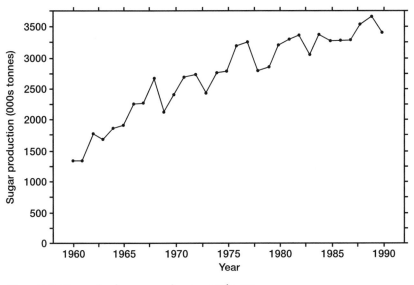

Figure 7.2 Queensland sugar production 1960–90
Source: Queensland Sugar Corporation (1991b)

and yet more intensification were by now, necessarily, limited by the fact that the existing production system was already highly intensive. What did occur during this period was an increase in the proportion of assigned land actually harvested each year (see figure 7.5). The very high percentage harvested in some periods may well reflect a tendency to extended ratooning, certainly figures over 80 per cent are not compatible with an optimum ratoon length of four or five years. In practice, the two phases of extended ratooning indicated in the graph seem to have occurred for two very different reasons. The high percentage of land being harvested during the early 1970s probably reflects attempts to capitalise on high sugar prices by deferring cultivation. Conversely, the similar situation which occurred during the 1980s, appears to reflect attempts to reduce cultivation costs during a period of very low returns.

World market sugar prices reached record high levels during the mid-1970s rising to around US$1,800 per tonne in 1974, and although they fell back to around US$600 per tonne in 1977–78 they had climbed again to over US$800 per tonne by the end of the decade. These relatively high prices underpinned a period of prosperity for the Australian Sugar industry, but this was short lived. Severely depressed prices which began in the early 1980s persisted throughout the decade and into the 1990s. The 'spot' price for world market sugar which had stood at US25 cents per pound in 1980 fell to as low as US2.8 cents per pound in 1985. Throughout the 1980s, the world market price for raw sugar averaged around US10 cents per pound.

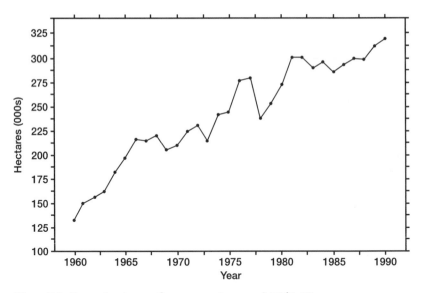

Figure 7.3 Queensland area of sugar cane harvested 1960–90

Source: Queensland Sugar Corporation (1991b)

Notwithstanding the effects of Australia's protected domestic market and long-term export contracts, the returns received by the Australian sugar industry were similarly around US10 cents per pound throughout this period. This level of return was often less than the costs of production in Australia which has been estimated to range from A$10.96 cents to A$13.25 cents per pound or roughly US8 cents to US10 cents per pound (ABARE, 1985:5). Thus cane farmers were faced with a classic 'cost-price squeeze' which, not unusually, tended to produce what might well be considered 'perverse' responses.

The exact numbers of farmers who experienced financial difficulties during this period is difficult to determine, but cash flow problems were certainly common. The average profitability of Queensland cane farms which peaked at A$75,000 per annum in 1980, fell to only A$12,000 in 1985. And it is clear that within this average large numbers of farms rapidly became sub-economic during the early years of the 1980s (Hungerford, 1987:82). A later survey undertaken by Gray *et al.* in 1992, indicated that around 60 per cent of Central Queensland farms (over 80 per cent of which were cane farms) were sub-economic at that time with the mean farm income standing at minus A$18,000. Less than 8 per cent of Central Queensland farms had incomes over A$40,000 in 1992 (Gray *et al.*, 1993:41).

Although the low sugar price did produce a moderate increase in the number of cane farms being sold during this period, the level of such sales is a poor indicator of the viability of the agricultural sector. A significant

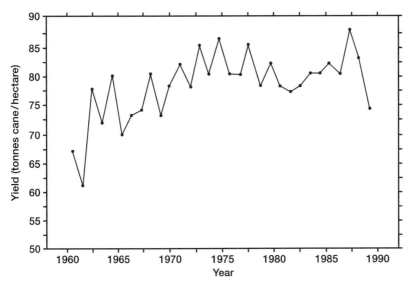

Figure 7.4 Queensland tonnes of cane per hectare 1960–90
Source: Queensland Sugar Corporation (1991b)

number of farmers were caught in a negative equity situation and were thus unable to sell their properties. In practice, statutory controls on the prices at which farms might be sold allied to a sustained period of low farm incomes meant that many properties were virtually unsaleable.

The Australian sugar industry regulatory system

For 70 years, almost every aspect of the Australian sugar industry was highly regulated. Statutory controls have covered not only the amount and location of land on which cane could be grown, but also whether or not that land might be sold and at what price. Farmers were also obliged to deliver their cane to a particular mill, and the framework for determining the price they would be paid was set out in legislation. The domestic market was protected and prices were fixed. Compulsory acquisition powers covering all sugar production also underpinned a system of centralised marketing whereby all Australian sugar was sold through the state governments or their agents.

The basis of the regulatory system existing in the late 1980s, evolved during the first 30 years of the twentieth century. Import controls were first enacted at the time of federation in 1901 and subsequently extended during the First World War. The 1915 Sugar Acquisition Act led to the fixing of the domestic retail sugar price and the granting of monopoly acquisition powers to the Queensland government. In practice, the fixing of retail prices led directly to the fixing of the prices paid to sugar producers and to

159

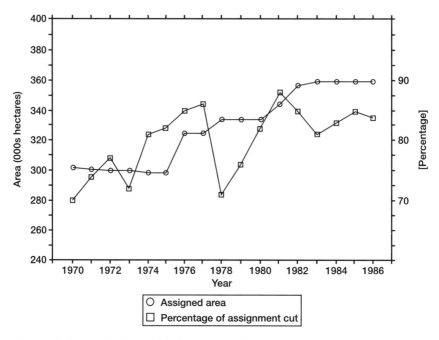

Figure 7.5 Queensland area of assignments and percentage cut
Source: Sugar Board (1991)

arrangements formalising the relationship between growers and millers. Further wide ranging controls over what land could be used for growing sugar cane were introduced during the 1930s.

The original rationale for regulation lay in what was seen to be the unequal relationship between millers and growers. Specifically, the dependence of farmers on particular mills, which created conditions in which the potential, at least, for exploitation clearly existed. But as one official enquiry into the industry pointed out, attempts to regulate this relationship led almost inevitably to a much more comprehensive regulatory system:

> Regulation of the sugar industry has a long history in Queensland, dating back to the early years of this century when small-scale cane growers were seen to be 'exploited' by the milling interests. Over the years, regulations were expanded and extended in response to a variety of circumstances, often to overcome the adverse incentives and 'loopholes' created by the original controls. In time, a large number of major economic decisions have become subject to approval from some regulatory body or another. These regulations have become an integral part of how the industry is organised and

operates, and a large section of the grower community is attached to them. Indeed, many find it difficult to envisage how the industry would operate in the absence of the rules and guidelines for behaviour provided by the existing regulations.

(Sugar Industry Commission, 1992:40)

In practice, although the regulatory system which evolved was extremely comprehensive and in some ways detailed and complex, the basic instruments used were relatively straightforward. Essentially, these involved: (i) protection of the domestic market; (ii) the designation of production quotas to different sectors of the industry; and (iii) powers to control exports and to determine the returns received by producers.

Protection of the domestic market   The Sugar Agreement Act (1923) established an embargo on the import of sugar into Australia. Subsequent to that date, all domestic demand for sugar has been met from Australian production at prices set by the Commonwealth government. Although the domestic price of sugar never included any great premium, this arrangement had several advantages for the Australian sugar industry. It provided a guaranteed market for a substantial proportion of total production, around 20 per cent in the 1980s. And perhaps even more significantly, because the incomes from all sugar sales were pooled before payments were made to processors and producers, the relatively stable prices received for domestic sales served to moderate the impact of often violent fluctuations in the world market price.

Peaks   The principal control on the supply side of the industry was known as a 'mill peak'. Each mill was granted a mill peak or quota, which represented a theoretical limit on the amount of sugar that mill could produce. In turn, each cane farmer using that particular mill was allocated a share of the mill's overall quota which was known as a farm peak. Designation of farm peaks effectively constituted a contract between growers and millers with the mill being obliged to buy the cane produced at a price established by a legislatively defined formula. The peak system functioned effectively because all sugar produced had to be sold into the state marketing system and production over peak would potentially, at least, receive prohibitively low prices.

Assignments   With the amount of sugar which a mill could produce being strictly controlled, it became necessary to control the number of growers who could supply that mill. With only a finite amount of production allowed, or more accurately only a finite amount of income available to the mill, any new farmers would inevitably prejudice the viability of existing producers. Such a scenario was avoided by restricting sugar cane agriculture to land specifically designated, or assigned, for that purpose. Such a system was enforceable because mills would only collect cane from assigned land.

Assignment could not normally be transferred from one piece of land to another. Because of this, much of the value of agricultural land existed in its assignment, rather than in the land *per se*. Consequently, further controls were established whereby assigned land could only be sold with the permission of the appropriate regulatory body who would determine not only whether a transaction could take place but also the price of the land.

Although mill peaks were periodically increased and the total amount of assigned land was adjusted in accordance with these, the established pattern of assignments was seldom modified significantly. Increased mill peaks were traditionally divided amongst established assignment holders on a *pro rata* basis. Thus, apart from rare occasions when, for example, grants of newly assigned land were made to ex-servicemen, the only feasible way to become a cane farmer was to purchase land which held an assignment.

Pools Pools were so named because all sugar produced in Queensland was pooled and marketed through a single agency with millers and, in turn, farmers receiving a *pro rata* share of the overall income. Payments for cane produced within peaks were calculated according to a formula not dissimilar in its operation to the 'A' and 'B' quota system operated with the EU (see for example Coote, 1987). In theory, the size of the number one pool was calculated to reflect the size of the domestic market and the amount of sugar being exported under contractual agreements. Thus producers could expect some degree of stability in the prices which they received for sugar sold within this pool. Conversely, the number two pool which would be sold on the world market would normally provide returns closely related to the prevailing world market price.

Somewhat incongruously, the prices paid for number two pool sugar were frequently not significantly lower than those paid for number one pool sugar. Indeed on occasions when the sugar market was particularly tight or when a sudden increase in the world market price of sugar occurred, they could exceed those of the first pool. Moreover, given that a high proportion of farmers' costs were fixed, production in excess of peak was often a rational strategy. In practice, production exceeded mill peaks in every year during the 1980s despite the fact that world market prices, and consequently number two pool prices, were depressed throughout most of the decade (Sugar Board, 1991:39).

Pooling arrangements have served to reduce some aspects of variability and uncertainty in the prices received by Australian sugar producers in two ways. First, all producers received a share of a single pool, so each producer received a mean price for the whole crop marketed that season irrespective of the particular market into which their own sugar was sold. Second, and related to this, producers received a price which related to sugar sold over the entire year. This is significant because world market sugar prices can change very rapidly and vary greatly within a single year.

In practice, for all the control exercised over the marketing of Australian sugar, the volatility of the world sugar market has been the key determinant of sugar industry incomes. Long-term export agreements have provided relatively secure markets, but these agreements were negotiated within a context of structural overproduction and never included any great premium over prevailing market prices. The only significant protection from the depressed and unstable prices occurring in the international sugar economy came from controls on the domestic market and these only served to alleviate the most extreme variability in returns from exports. Since 1970, total sugar industry receipts have closely reflected the world market price for raw sugar which varied by as much as 73 per cent from one year to the next. Because the prices actually paid to the agricultural and milling sectors were a direct function of total industry receipts, the gross incomes received by both growers and millers varied at the same rates as the total industry income throughout this period (Powell and McGovern, 1987).

The overall regulatory regime functioned effectively because of the total interdependence of different sectors of the industry. This allowed a very high degree of control to be achieved through the use of a limited number of measures. For example, by controlling the output and incomes of the mills it was possible to delegate more detailed regulatory functions to this level. Thus actual government involvement in the day to day operation of the industry was never that great. The Commonwealth government's role was largely confined to strategic considerations such as international terms of trade. Most statutory controls on the sugar industry, for instance the designation of peaks and those affecting the single body marketing of sugar, have been enacted at state level, but in practice their operation has normally been delegated to various industry bodies such as milling companies. However, a number of governmental, quasi-governmental and producer organisations have been directly involved in the regulation of the Queensland sugar industry. The most significant of these are:

The Department of Primary Industries (DPI)   The DPI is directly involved with the sugar industry in a number of areas. The DPI's principal area of responsibility is in environmental and water management – a remit which includes responsibility for industry specific issues such as nutrient leaching, cane burning, irrigation and water extraction policy. The DPI also has a responsibility for promoting efficient farm management, which according to the DPI involves 'planning for sustainability and profit' (DPI, 1994). This department also administers the Rural Adjustment Scheme and other initiatives to help farmers experiencing economic problems.

Cane protection and prices boards   The Central Sugar Cane Prices Board was established by the Queensland government in 1915. Although the initial role of the Board involved ensuring an equitable distribution of income between the growers and millers, its remit was progressively expanded until it controlled virtually every aspect of sugar production and marketing. In 1933, the Board became responsible for controlling the area and distribution of land on which sugar cane might be grown. Allied to this it also controlled the price at which any assigned land might be sold. Indeed no assigned land could be sold or transferred without the Board's authority. In 1939, the Board undertook responsibility for determining and annually reviewing mill peaks. In theory, this Board also had responsibility for the acquisition and single desk marketing of all the sugar produced in Queensland. In practice, the administration of the marketing system has been devolved to CSR Ltd.

Local sugar cane prices boards Every operating mill had a local sugar cane prices board which undertook detailed regulatory functions at a local level. Local boards met each year to determine the farm peaks, i.e. the way in which the mill peaks were divided amongst the farmers assigned to that mill. Local boards also administered the division of receipts between the milling and agricultural sectors and amongst individual farmers.

The Bureau of Sugar Experiment Stations The Bureau of Sugar Experiment Stations (BSES) is the principal research, development and extension agency operating within the Australian sugar industry. The BSES develops new cane varieties, has responsibility for disease control, and evaluates new technologies within the industry. The Bureau is the largest provider of extension services to sugar cane farmers (BSES, 1992).

Canegrowers Although funded by a statutory levy on sugar production, Canegrowers is essentially a producer organisation which represents the interests of cane farmers. All cane farmers are members of Canegrowers which undertakes a number of responsibilities on their behalf. At a strategic level, Canegrowers lobbies government and negotiates with large milling interests. At a local level Canegrowers also provides a range of extension services and will undertake a range of administrative functions such as bookkeeping for those farmers who choose to use these facilities.

Other institutions A number of other organisations, such as the Australian Sugar Milling Council, the Sugar Research Institute and the Sugar Research and Development Corporation, are also involved in the sugar industry. A less industry specific, but nevertheless highly significant role was also played by the banking sector. Following the privatisation of the state Agricultural Development Bank in the 1970s, neither Commonwealth nor the state governments have had any great involvement in the financing of the sugar industry. Virtually all capital requirements have been financed through private banks.

Deregulation

Deregulation of the agricultural sector has been part of wider Australian government policy for some years (Alston, 1991; Lawrence *et al.*, 1992). Australia's near neighbour New Zealand has pursued similar policies (Cloke and Le Heron, 1994). The sugar industry was the last major sector of Australian agriculture to undergo deregulation. However, by the late 1980s a comprehensive programme of deregulation of the sugar industry was being instituted. Given the extremely high level of regulation which had previously pertained, deregulation has involved profound and potentially highly significant transformations.

One of the initial steps in the deregulatory process was the discontinuance in 1988 of the embargo on sugar imports, which had existed since the 1920s. This was replaced with a system of import tariffs which are to be progressively reduced. Even when the last tariff protection is eliminated in the late 1990s, this is unlikely to produce any importation of sugar into Australia because of the transport costs involved. However, the concurrent termination of the arrangements covering the supply and pricing of the domestic market could be more significant. In theory, this might produce a competitive market with a consequent reduction in the price paid to producers. Certainly the Sugar Industry Commission (1992) suggested that domestic consumers were likely to benefit from price reductions. In practice, with one refiner supplying well over 90 per cent of the domestic market, significant retail price reductions are unlikely to occur. Of more concern to producers is the prospect of CSR being in a position to play off one supplier against another and thus reduce returns to the milling and agricultural sectors.

In 1991, the Queensland government passed the Sugar Industry Act which effected a major revision of the legislation and associated administrative arrangements affecting the Queensland sugar industry. One major outcome of the Act was the formation of the Queensland Sugar Corporation which absorbed the marketing responsibilities of the Sugar Board and the regulatory functions of the Central Sugar Cane Prices Board. The remit of the Corporation includes: 'development and implementation of policy relating to the management of the Queensland sugar industry; managing and regulating the quantity and quality of sugar cane and raw sugar produced in Queensland; marketing each season's raw sugar and distributing the proceeds from the sales of raw sugar' (Queensland Sugar Corporation, 1992b:42). Under the provisions of the Sugar Industry Act, the assignment and peak systems remained essentially intact, although some of the more far-reaching controls, such as those relating to the sale of assigned land, were discontinued. Determination and administration of peaks and assignments became the responsibility of the Queensland Sugar Corporation who extended the total amount of assigned land by 8 per cent in 1991, and scheduled further increases of 2.5 per cent for each year up to 1995.

In March 1991, some years after the deregulatory process had been initiated, the Australian government ordered the Sugar Industry Commission to consider the future of the sugar industry. The terms of reference given the Commission included the requirement that they should 'review production, institutional, regulatory or other arrangements subject to influence by the Governments in Australia . . . and identify any further initiatives which will raise overall economic efficiency' (Sugar Industry Commission, 1992:xiv). The Commission's report published in 1992 argued strongly in favour of continuing and extending the deregulatory programme. It cited several arguments to support this conclusion. These included the suggestion that both the international context in which sugar was exported and the nature of the industry itself had changed significantly since the regulatory framework had been established, and the fact that regulations applying to other sectors of Australian agriculture had already been relaxed or removed.

The Commission also argued that the regulatory system had imposed a number of specific costs on the industry and the Australian economy. It was suggested that the expansion of the industry had been inappropriately constrained and that potential exports had been lost. The assignment system had produced situations where cane was grown in sub-optimal locations while more appropriate, potentially more productive, land was not utilised. Restrictions on the transfer of assignment had prevented a restructuring of the industry in which the scale of production units would have increased to take advantage of potential economies of scale. The constraints of the regulatory system were also claimed to have stifled innovation amongst growers and promoted environmentally unsound practices including the overuse of chemical inputs and inappropriate reductions in fallowing. For example, farmers who were constrained from increasing their holdings by the assignment system, were often left with little alternative but to attempt to intensify production on their existing land. It was also suggested that controls on marketing had imposed a cost on Australian sugar consumers of around A$1,000m per year (Sugar Industry Commission, 1992).

The Sugar Industry Commission proposed a range of further deregulatory measures which they suggested would represent 'a major departure from current arrangements' (Sugar Industry Commission, 1992:6). These recommendations included, removing all remaining administrative constraints on the area and location of cane growing; allowing growers and millers to negotiate the terms of their relationship themselves; the development of competition between multiple sellers of sugar on the domestic market; and the development of new arrangements for pooling and marketing sugar. In the Commission's words these changes would enable 'growers, millers and marketers to evaluate alternative strategies and enter into those arrangements which best suit their individual needs. Competitive pressures would provide a strong incentive to ensure that the production and marketing activities are undertaken as efficiently as possible' (Sugar Industry Commission, 1992:6).

Most of the deregulatory measures proposed by the Commission are either currently being put into practice, or are planned for the near future. The only real area of uncertainty regarding the extent of the deregulatory process lies in the question of whether the single desk marketing arrangements for Australian sugar continue beyond the short term. The Commonwealth government appears to favour a total deregulation of the industry including the abandonment of the unitary marketing arrangements. There are, however, strong arguments that the existing arrangements for single desk marketing or something effectively very similar should be maintained (Queensland Sugar Corporation, 1992c).

Although there is general support for some elements of the current deregulatory programme, in particular for the abolition of many of the detailed controls over what are seen as trivial and insignificant features of the industry, both farmers and the milling sector are aware that deregulation is likely to create new pressures and problems within the industry. Various actors within the industry tend to be both sceptical of the reasons for deregulation and concerned about the possible implications of deregulation for them as individuals. Farmers, in particular, see deregulation, or more precisely the scale of the proposed deregulatory agenda, as being ill-conceived given the fact that the established Australian sugar industry is widely perceived to be amongst the most productive and efficient in the world. This lack of enthusiasm for deregulation was noted by the Industry Commission who acknowledged that, 'it is easy to understand why many in the industry supported regulation' (Sugar Industry Commission, 1992:2). One widely held view is that the deregulation of the industry simply reflects the ill-considered application of an increasingly neo-liberal political agenda to Australian agriculture. Another point of view is that the Australian government has effectively backed itself into a corner through its criticism of other sugar producers. Certainly, the Australian government has been extremely vocal, for example as a leading member of the 'Cairns Group', in its advocacy of reforms to international trading arrangements and its criticism of the protectionist policies pursued by many foreign governments, including both Europe and Japan (Miller, 1987; Jesson, 1991). The suggestion being that because of the stance which the Commonwealth government has taken over these issues, it has felt obliged to 'put its own house in order'. In effect, it has been obliged to dismantle the tight regulatory framework within which the sugar industry had operated.

Case studies: Bundaberg and Mackay

The early 1990s may well prove to have been a highly significant period for the Australian sugar industry. The 1980s had seen a sustained period of depressed sugar prices, during which almost all cane farmers experienced financial difficulties and many ceased to produce cane. By 1994, incomes were beginning to rise as small increases were occurring in world market sugar prices, but a range of problems had been accentuated and brought to the fore by the conditions which prevailed during the previous decade. Moreover, uncertainty about the scale and nature of the deregulatory process and the effects which it is likely to produce have become a major cause for concern. Whether the deregulatory process will produce a more efficient and competitive industry remains to be seen, but irrespective of this, it seems inevitable that there will be both winners and losers as the changed context within which the industry operates induces a period of restructuring.

Bundaberg and Mackay are two of the major sugar producing regions in Queensland. They are in many respects typical of the sugar producing regions found throughout the state. Certainly, the farmers in these two regions have experienced many of the same problems which have been common throughout the industry over recent years. Equally, however, both of these regions have a number of unique features. For example, the Milling sector in Bundaberg is controlled by a private company whereas in Mackay it operates as a co-operative. There have also been significant differences in the coping strategies adopted by the cane farmers in the two locations.

The first cane sugar produced in the Bundaberg region was grown on the banks of the Burnett River in 1882. By 1900, a significant industry had developed around a number of large centralised mills. The local industry continued to expand throughout the twentieth century. By 1991, 47,934 hectares of land were assigned to sugar cane and the region held a peak of 400,400 tonnes of sugar, 12.6 per cent of total Queensland production. In 1992, there were approximately 830 individual cane farmers in the Bundaberg region, with the average size of individual holdings standing at just over 60 hectares (Bundaberg Canegrowers, 1991:2). Many local farmers experienced severe difficulties during the 1980s when sugar prices remained consistently depressed. Cane production which had stood at over 3.1 million tonnes of cane in the early 1980s had fallen to just over 2.5 million tonnes by the early 1990s, by which time sugar production was approximately 7 per cent below peak (ABARE, 1991).

Fairymead, the first sugar milling company in the Bundaberg district, was established in 1880. Several other sugar milling companies also came into existence during the 1880s. Initially the Fairymead company had quite diverse agricultural interests, including cattle production, but sugar and related activities soon became its central concern. Fairymead became a public company in 1912 and was to play an increasingly central role in the development of the local sugar industry. Fairymead changed its name to the Bundaberg Sugar Company Ltd in 1972 after merging with another local sugar company. By the 1980s, Bundaberg Sugar owned all of the sugar mills in the Bundaberg area, was a major rum producer and Australia's second largest sugar refiner. Although the majority of the five million tonnes of sugar cane produced in the Bundaberg area each year is grown on relatively small family farms, Bundaberg Sugar is almost unique in Australia in that the company owns and farms estates amounting to over 7,000 hectares of cane land in its own right.

In 1991, Bundaberg Sugar was acquired by Tate and Lyle plc, a British sugar multinational. Such a development would not have been possible prior to this date because of a statutory ban on any foreign ownership of assets in the sugar industry. Tate and Lyle's takeover of Bundaberg Sugar was received with some considerable misgivings by many of those involved in the local sugar industry. Local cane farmers, in particular, have been concerned about the implications which the takeover might have for them. Although Tate and Lyle are a major sugar multinational with interests throughout much of the world, the reasons for their interest in Bundaberg Sugar are somewhat unclear. It may be that Tate and Lyle have some interest in gaining a foothold in the supply of refined sugar to the domestic Australian market, but this is a relatively small market and Bundaberg sugar has a much smaller market share than CSR. It would appear more likely that Tate and Lyle see Australia as a possible entry point into the supposedly rapidly developing Asian market for refined sugar. Alternatively, many sugar cane farmers and millers in Queensland fear that foreign companies such as Tate and Lyle may simply be involved in an asset stripping exercise. Bundaberg Sugar, and several of the other milling companies in Queensland are relatively asset rich and it may well be that a company with the inclination and resources could make considerable short-term gains in this way. In practice, however, since becoming established in Queensland, Tate and Lyle have attempted to expand their involvement through the acquisition of other milling interests in the state. This has proved to be a difficult process as they have encountered quite considerable opposition from various sectors of the industry. For example, the members of at least one sugar co-operative in northern Queensland voted overwhelmingly to reject a takeover in 1994.

Sugar cane agriculture in the Mackay area extends for 50 or so kilometres inland along the valley of the Pioneer River. As is the case with most sugar producing regions in Queensland, sugar production began here in the second half of the nineteenth century and has continued to expand since then. In the early 1990s there were approximately 1,200 cane farmers producing cane on approximately 125,000 hectares of assigned land in the Mackay region (ABARE, 1991:24). In 1991, the Mackay region produced over 600,000 tonnes of raw sugar which amounted to around 17 per cent of total Australian production. However, this level of production was somewhat lower than had been achieved in previous years. In some areas, production had fallen by as much as 40 per cent compared to levels achieved in the late 1980s (Mackay Canegrowers, 1994). A proportion of this decline may have been due to a number of low rainfall years, but the financial difficulties experienced by local farmers during the 1980s also appears to have been significant.

Sugar milling in this region is undertaken by the Mackay Co-operative of which all local growers are members. The present company was founded in 1988 when the five separate co-operative sugar milling operations which then existed were consolidated and rationalised (Kerr, 1988). The co-operative now operates four large mills which employ around 1,000 full-time workers (Mackay Sugar, 1994). During 1994, the Mackay Co-operative was actively considering changing its status into that of a public company. The principle reason for this being that as a co-operative they face a number of technical problems in raising the capital which they believe is needed to finance future development. By this time, the co-operative was already engaged in a joint venture with the British multinational R. D. and F. Man to develop refined sugar exports from Australia.

Technological treadmill

The Australian sugar industry has to compete with a number of other sugar producing countries within the world sugar market. Many of these producers, such as Thailand, Cuba and Brazil, are low wage economies. Others such as the EU give their sugar producers significant amounts of protection and support. Australian wages are amongst the highest in the world and sugar producers receive little if any direct support from the government. Given this context, the Australian industry sees its one comparative advantage in being able to develop and adopt new technology more readily than other sugar exporting countries. Certainly, this seems to be the position of the Queensland government:

> The responsibility is your own. If you want to be a better farmer you have to be determined, the tools are there to assist and help you. We do our best in the research area. The BSES and other research groups have the responsibility to help to bring in modern and better farming systems and also cane varieties that can give you better productivity. That role has really not changed since the early days of the BSES, but what has changed is the technology and the ways in which it can be implemented. There is not one of you who is operating on the farm in the same way as did your parents in time gone by because you must move with the times. . . . The fact that we will have over 4 million tonnes of sugar this year has been because of proper planning and new legislation that the government has promoted in conjunction with the industry. That is what it is all about today, the new technology you see here today is all designed to help you lower your cost structure in some way or another. Because margins are shrinking – they are nowhere near what they were years ago. Because we are growing for overseas consumers our cost structures are for ever and a day going to be determined by those overseas prices . . . the alternative is that four out of every five cane farmers would have to go out of

the industry. Our package is designed to lower that cost structure. One of the biggest problems for farmers is that you are constantly adapting to change, to changes in other areas. But by adapting your farm you also ensure that you are adapting your livelihoods. So we must keep up the research and extension work. . . . You have got to have the will, the understanding and the inclination to want to move yourself down a better track.

(Queensland Minister of Primary Industries, Speech to
Mackay cane farmers, April, 1994)

Whatever the political rhetoric, however, it is clear enough that constant pressure to improve efficiency, to adopt new production techniques, to use more modern technology, is always likely to cause problems. Certainly, it has had profound and sometimes overwhelming implications for the agricultural sector. Pressure to modernise and expand production tends to cause unsustainability because of the economic costs involved, but it can also prejudice the sustainability of farming enterprises in other ways. Modernisation frequently involves more than the mere purchase of new machinery or the adoption of new production techniques. In practice, it may well mean that a farmer has to acquire new assignments if these investments are to be used efficiently. Certainly the pressures to increase farm size appear clear enough to many of those working in the industry:

If it cost $24,000 per year personal expenses, for someone growing 3,000 tonnes which is about the average Millaquin cane farm, that means that they have got to take $8 per tonne profit to live. We are growing 8,000 tonnes and it still only costs us $24,000 to live, so out of the crop we only have to take $3 per tonne. So when the guy that is growing 3,000 tonnes is breaking even, we've got $5 per tonne left and on 8,000 tonnes and that is $40,000. When the other bloke is breaking even we've got $40,000 profit.

(Personal communication, Bundaberg cane farmer)

Queensland cane farmers have also been directly affected by modernisation of the milling sector. During the early 1990s, many Queensland mills have been transforming their operations from the traditional pattern of daytime only working to what is known as 'continuous crushing'. Traditionally, cane cutting stopped in the early evening and neither the farms nor the mills worked during the night. Under a continuous crushing system, cane must be harvested during the evening to ensure the mills with adequate supplies throughout the night which is technically much more efficient. The nature of sugar cane production is such that if a factory changes to continuous crushing, all the farms in its area must adopt new harvesting practices. In practice, extending the working day from early morning to late evening can create very real difficulties for a small family farm which relies exclusively on family labour.

The modernist, technocentric development ideology which has shaped the development of the Australian sugar industry is in many ways analogous to the reliance on progressive technical innovation which forms a central tenet of neo-liberal thinking on sustainable development. For at least three reasons, it is far from clear whether this sort of development paradigm can form the basis of a truly sustainable system in practice. First, as the situation in Australia demonstrates very clearly, agriculture, and certainly sugar cane agriculture, is inextricably related to nature. Thus however efficient production is made, there are absolute limits to productivity. Although whether or not the Australian industry has actually reached these limits is questionable, it is clear enough that incremental efficiency gains become progressively harder to achieve once an industry has reached a level of intensification such as that already practised in Australia. Second, in a situation such as this, it is always likely to be the case that further efficiency gains will be achieved through social or environmental exploitation. And, third, the context in which this sort of 'development' occurs is dynamic and competitive and any efficiency gains which are achieved will never be sufficient to make production economically sustainable in more than the short term. Not least because many of Australia's international competitors now have both low wage economies and access to increasingly sophisticated production technologies.

Potential and actual unsustainability

Queensland cane farmers operate in a context where various forms of risk and uncertainty are omnipresent and highly significant, but what has been occurring in the sugar sector in recent years reflects more than the kind of short-term problems caused by fluctuations in the sugar price or the weather. The context within which the industry operates has become progressively more stressful. On the one hand, the sugar price, volatility aside, has been declining in real terms for several decades. In parallel with this, the current structure of the industry has also been prejudiced by a range of internally generated contradictions including those which are undermining the viability of the family farm. By the late 1980s, it had become increasingly unlikely that either the existing regulatory framework or the sort of incremental technological advances which had sustained the established industry structure for almost a century would be sufficient to sustain it into the future. Indeed, these regulatory arrangements are now being disbanded and the Australian sugar industry is, apparently, entering a period of profound restructuring.

Whilst many of the pressures faced were common to both of the case study locations and indeed throughout the industry as a whole, some problems were more place specific, as were many of the responses adopted. Indeed both pressures and responses differed quite markedly not only between one sugar producing region and another, but also within individual regions. In practice, while the nature of sugar cane agriculture and a highly prescriptive regulatory system produced many elements of commonality amongst producers, significant differences remained in both the practices and the productivity of both cane farmers and the millers. As one study of the industry pointed out:

> While cane growing occurs on a large number of small farms, there is considerable diversity both in the mode and efficiency of operation as well as in the effective return obtained. This is due to both general factors such as weather and seasonality, and also to individual features of the farmer, his operation (for example, the use of contractors or his own labour and plant) and his land. Performance does vary to a great, and probably unexpected, extent. ... Similarly, there is a variation in mill situations and performances. While behaviours and modes of operation differ little between mills (there is a generally uniform use of technology and techniques), their financial position and state of development differ.
>
> (Powell and McGovern, 1987:9)

The situation is not only more diverse than sometimes believed, but also more complex. Certainly, individual farming enterprises have tended to become economically unsustainable as a result of a combination of factors, although these are often interrelated. As one DPI official suggested:

> Policy makers always like to look for one explanation, perhaps to say that it was dry land farmers with poor management experience. But financial problems can come from things like family break-up at the wrong time; purchasing a property at the wrong time and taking on a great deal of debt; or poor management. And it's never just one, it's usually a combination. Locally, there were particular areas which didn't get rain. Some growers were preoccupied with taxation and had absolutely everything on lease so the level of fixed commitments going out was ridiculous. Another problem is with succession, where they are trying to get their son or sons onto the farm and having too many families reliant on too small a farm.
>
> (Personal communication, DPI Farm Financial Advisor, Mackay)

Agronomic and environmental problems

Although a range of agronomic problems have been evident in Australian sugar cane agriculture, few farmers in either of the case study locations or indeed elsewhere have become unsustainable simply because of these. As is almost inevitable given the monocultural nature of sugar cane agriculture, farmers in Queensland have periodically faced a range of disease problems. A good example of this is the relatively widespread problem of ratoon stunting disease which affects the regrowth of ratoons (BSES, 1992:17). The usual response to disease problems in sugar cane is the development of new disease resistant varieties. Research and development including cane breeding and related extension services are well developed in Australia and the majority of farmers appear to be well satisfied with the manner in which new varieties are developed and disseminated. Whilst it is generally recognised that intensive monoculture is an inherently problematic form of agriculture (Buttel and Gertler, 1982; Cameron and Elix, 1991; Burch et al., 1992; Hindmarsh, 1992), there is a general belief in Queensland that any problems which do arise can be managed. The assignment system and the industry culture mean that few farmers have ever seriously considered deviating from traditional monocultural production techniques. When new practices have been adopted this has usually been done reluctantly and very much as a last option. For example, a number of Bundaberg farmers were more or less obliged to diversify during the 1980s, but this diversification generally reflected the pressures created by cash flow problems rather than any perceived advantages in a more diversified form of agriculture.

Drought is the one environmental constraint which affects significant numbers of Australian cane farmers. Rainfall patterns are highly unpredictable throughout the whole of Queensland and drought or variations in seasonality can have a significant impact on yields, particularly where irrigation is not used. The Mackay region, for example, was quite severely affected by a series of low rainfall years in the late 1980s and early 1990s. Although irrigation reduces some risks, it is not without its problems. In most cases charges for water are far from insignificant. In some areas, including both the case study locations, there are also potentially significant problems of salt water intrusion into aquifers. The semi-arid climate and lateritic soils of central and southern Queensland are also such that salinisation can be a problem, although only moderate amounts of land have been severely affected to date.

Bundaberg is one of the drier sugar producing areas in Queensland with average rainfall varying from 1,039 mm in the western sector to 1,114 mm per annum in the east. Over 95 per cent of the land in the district is irrigated. Land along the Burnett river is normally irrigated using water from that source. Most farmers who do not have access to the river have traditionally pumped water from an aquifer which underlies much of the region. A series of low rainfall years during the 1960s and concerns about overextraction from the aquifer prompted the development of the Bundaberg–Isis irrigation scheme work on which commenced in 1970. One local cane farmer explained the need for this scheme:

Nearer the river it started to get salty, and it was gradually spreading this way. So if we hadn't gone on to surface water from the dam, there would be no irrigating at all. It was a bit too late for some down that end, but here, some of them growled that they still had good water, but the salt was spreading; it was getting further and further. You only had to get a dry year and we would get it here. They just over-watered. Once the salt got in there was nothing to do. Now, we pay quite a lot for it, but you know that the quality of the water is good.

(Personal communication)

This was a major scheme intended to provide irrigation water throughout most of the Bundaberg district. However, progress in construction has been slow and the scheme remains incomplete, not least because successive governments have equivocated over the costs involved (Hungerford, 1987). Those elements of the irrigation scheme which are operational have allowed more farmers to irrigate and this has reduced pressure on the aquifer, but current levels of groundwater extraction still remain above replenishment rates in some areas.

As with most aspects of sugar cane agriculture, water extraction rights and charges have been subject to a considerable degree of regulation. Farmers are allocated a quota of water according to the size of their assignment. This basic quota must be paid for irrespective of whether it is used and extraction beyond this quota is charged at prohibitive rates. Historically, most cane farmers used flood irrigation techniques, however, a transfer to drip irrigation is currently being strongly promoted by both the DPI and other regulatory bodies. Certainly a significant number of farmers in both Bundaberg and Mackay are currently adopting this new technology. One Bundaberg farmer was clearly enthusiastic:

Sugar production can only be made profitable through technology. We are installing trickle right through the farm . . . there are large areas of land suitable for trickle here and an extra million tonnes of cane is achievable in the three Bundaberg mill areas. It is a very attractive commercial proposition, there is an enormous financial attraction in the $45 million of extra revenue this would produce . . . some people dither because they believe that the technology may involve some pitfalls, but there are none, it's out there for them to see. The only problem is that too many farms are too small to finance installation.

(Personal communication)

A move to drip irrigation may appear rational in that it reduces pressure on scarce water resources. In practice, however, pressure to adopt such technology has been a significant causal factor underlying the unsustainability of many farming enterprises. The adoption of drip forces inefficiencies upon those who cannot adopt and a very real threat of debt crisis on those that do. As a more reflective cane farmer commented:

> At the moment in Bundaberg there is a thrust towards trickle irrigation. Now the capital inputs into trickle are enormous. And you really are taking a gamble, particularly in some areas. The running costs are cheap and the efficiency are good – you can run with low pumping costs and also introduce chemicals and nutrients right to the stool with no waste, but the capital outlays are enormous. They talk in terms of around $2,000 per acre. Well if you have got a 250 acre farm it's almost the value of the farm again. You have got to know that you can handle that. And some people do get into trouble. The guy who sits on the fence, he may put a patch in but carries on with the rest, he's the survivor.
>
> (Personal communication)

As this comment implies, the events of recent years may have given cane farmers good cause to be circumspect, but it still tends to be the speed of modernisation which is questioned rather than the process itself. This is perhaps somewhat surprising given that so many farmers now have unsustainable debts deriving more or less directly from inappropriate investments in technology. Moreover, although water policy is tending to move towards various forms of demand management, such as the adoption of more efficient irrigation techniques, such measures are still embedded within a positivist, modernist interpretation of development. Both the regulatory agencies and the farming community have an almost unquestioning faith in potential of technology to solve any agronomic or environmental problems which emerge (see for example, Bundaberg Cane Productivity Committee, 1993). Moreover, development strategies for the sugar industry are still formulated within a context defined by the fundamentally productionist ethos of wider agricultural policy (Wheelwright, 1990). In practice, the long established productionist philosophy of the Australian government and the nature of the sugar industry regulatory system have combined to produce increasingly intensive farming methods throughout the sugar industry. Farmers have been encouraged to increase production on finite amounts of land and this has, almost inevitably, led to high levels of fertiliser, pesticide and herbicide use. This trend has often been accentuated by the strategies adopted by many farmers in response to the cost–price squeeze situation such as occurred during the 1980s. When faced with low sugar prices, the initial reaction, at least, of many farmers appears to have been to attempt to further increase yields.

The highly intensive, chemical dependent, nature of Australian sugar cane agriculture is somewhat at odds with that country's widely promoted 'clean and green' image and an increasingly influential, and often government supported, environmentalist movement (see for example, Campbell, 1989; Cock, 1992). In practical terms however, pressure from green movements has had little direct impact on Queensland cane farmers or the milling industry. In part, this may reflect the very high dependence on this one industry in sugar producing regions such as Bundaberg and Mackay. The problems associated with cane burning – large amounts of smoke and ash deposits in adjacent areas, for example, are normally considered to be a minor and necessary inconvenience by populations whose livelihoods are dependent on the industry. The DPI is currently promoting a change to cutting green cane rather than burning fields before harvesting, but this is because green cane harvesting is now perceived to have agronomic advantages rather than because of pressure to change from outside the industry. The effects of nutrient leaching on the Great Barrier Reef are seen by both the green movement and the regulatory authorities as a potentially significant problem. Run-off of agricultural fertilisers and effluent from sugar mills are both considered to be major contributors to the overall problem. That said, there is little evidence that either the mills or the farmers have faced any real pressure to moderate their practices because of this. In practice, the key problem for most cane farmers has not been the environmental impacts of current practices, but rather the economic implications of operating highly capital intensive production techniques in a situation of volatile and declining sugar prices. The prevailing mode of social regulation has been such that the agricultural system's effects upon nature have remained largely hidden.

Unstable and uncertain incomes

Historically, incomes within the Australian sugar industry have always been both highly variable and highly unpredictable. Uncertain incomes are problematic for the milling sector who need to raise capital and subsequently finance borrowing in order to develop their operations. Milling co-operatives face particular difficulties here in that they are more constrained in the ways in which they can acquire capital than proprietary companies. The Mackay Co-operative, for example, sees the problem of being restricted to equity finance as a particular problem in a period of restructuring such as is now occurring because this makes them vulnerable to the expansionist designs of larger and better financed players (personal communication, Mackay Sugar executive).

A situation where incomes are insecure also creates extreme difficulties for individual cane farmers. Long-term planning is at best an uncertain exercise and this creates particular problems where farm development is necessarily a long-term exercise and, moreover, one which almost inevitably has to be financed over an extended period. In practice, even the most astute farmers tend to find that their actions are often determined more by unforeseen short-term pressures than any strategy for long-term development. In both Bundaberg and Mackay the long-term sustainability of many farming enterprises has often been prejudiced by short-term pressures. As one Bundaberg cane farmer suggested:

> You do your best to plan for the future, to develop your property. You listen to the experts and you do your best. You borrow money to stay up to date, to buy new blocks. You do what you are told and what seems right and then things change and you've got your back to the wall. Prices go right down and interest rates go right up. What do you do then? That is one of the worst parts of it. It wasn't because they didn't work. It makes them feel terrible because you can go on the dole and live on the beach, but these farmers, they worked hard, they worked long hours, they were good farmers. And you know you must get some thoughts that maybe it was something I've been doing. We've got neighbours just here, they worked really hard, they were good farmers and they haven't made a cent. It's not as if they've done the wrong thing, it's just the circumstances.
>
> (Personal communication)

Debt

The year on year variability of sugar industry incomes is such that even the most efficient cane farmers need to balance years with high returns against those when little if any profit is made. This volatility is a structural feature of the global sugar economy, and indeed one common to many agricultural commodities. The unusually long period of depressed prices and incomes which occurred during the 1980s, however, resulted in a very large proportion of Queensland's cane farmers experiencing a situation where they had little or no income for several years. Under these circumstances, it was not only the least efficient, least well managed farms which were threatened. In practice, large numbers of farms, which under less extreme circumstances would have been both highly productive and quite viable, were becoming financially unsustainable. A DPI Farm Financial Services Advisor in Bundaberg estimated that the department had been involved with around 10 per cent of local cane farms each year during the 1980s (personal communication). A similar situation also existed in Mackay:

> Personally, I was seeing around 125 farming families out of 1,500 in the central district. Between myself and Canegrowers we have probably seen at least 15 per cent of canegrowers. Others were actually seeing their accountants, some relied on business friends and other

178

more experienced cane growers. There is no doubt about it, the sugar industry in the Mackay district went through a critical period two years ago. One more year would have seen unbelievable consequences.

(Personal communication, Farm Financial Services
Manager, DPI, Mackay)

By 1992 well over half of Central Queensland farmers had debt to assets ratios of over 20 per cent, and in around a third of cases the ratio was over 40 per cent (Gray *et al.*, 1993:40). High levels of borrowing, low incomes and interest rates which rose to almost 20 per cent left many farmers with unsustainable debt burdens during the 1980s. Many were forced to sell their farms and even more were left in a situation where they could neither service their debts nor sell their properties either because they could not find a buyer or because the price available would not cover their debts.

In practice, farmers tended to have high levels of debt for one of several reasons: there were new entrants who had borrowed heavily to finance the purchase of a cane farm; there were those who had invested heavily in new machinery at the beginning of the 1980s; and there were those established farmers who had attempted to expand too fast by purchasing other farms. However, there seems to be little doubt that both the banks and the government bear some responsibility for the financial problems which developed within the agricultural sector. Both had positively and quite vehemently encouraged high levels of borrowing during the 1970 and early 1980s. On the one hand, advice from government consistently promoted both intensification and the expansion of individual holdings – 'get big or get out' had been the catch phrase of the 1970s. In parallel with this, however, the banks had not only encouraged farmers to borrow money, but in retrospect, they had clearly extended inappropriately high levels of credit to individual farmers. As one cane farmer suggested 'the only trouble we have with our bank is that whenever we want a loan he'll let us have it' (personal communication). A DPI Farm Financial Advisor commented on the position adopted by the banking sector in these terms:

I really wonder about the banks and I have to deal with them all the time. They tend to work very much on short-term criteria. But what else can you work on because time and time again longer-term predictions have been proved to be wrong.

(Personal communication)

Over the last few years, the financial sector has modified its lending criteria regarding loans to cane farmers; moving away from equity based determinations to a policy of evaluating loans on the basis of income generation potential. While they are generally highly critical of the banks' original lending policies, most farmers believe that the banks adopted reasonably sympathetic approaches towards indebted farmers. In practice, of course, the banks' options were then very limited given that a very high percentage of individual farmers had debts which exceeded their equity.

When asked why local farming enterprises had become unsustainable, a very high percentage of interviewees in both case study locations made a clear distinction between those cane farmers with no debts and those with significant debt burdens. The suggestion being that those with no debts could withstand periods of depressed incomes simply by postponing any major farm purchases and perhaps cutting back on their personal expenditure. Whereas those with high levels of debt had little opportunity of remaining solvent whatever type of strategy they chose to adopt. As one established cane farmer put it:

> I came into it in 1950 so I had 30 years when it was a boom industry. There was never any doubt that I would survive. But if anyone had done the same thing as I did in the '80s they wouldn't have survived. By the time it got to the '80s I owned everything and I didn't have a debt. If I had done exactly what I did in 1950 in 1980 I would have really had problems . . . there were a lot of them around here who bought farms for half a million dollars and they borrowed half of that. Within four years the farm was only worth a quarter of a million, the value of the farm had halved. They still owed a quarter of a million. They had no equity at all. Those circumstances drove a lot of people into growing small crops. They were clutching at straws because they were in such a bad position. . . . Debt was the biggest thing. I think so. They got in it when it dropped virtually 100 per cent in a couple of years and interest rates went from 12 per cent to 24 per cent. They didn't know it was going to happen, nobody did.
>
> (Personal communication)

In itself, the suggestion that farmers with little or no debt were better placed to withstand a prolonged period of low incomes appears to be an obvious truism. A key point, however, is that large debts usually reflected measures undertaken with the objective of making the farming enterprise more productive, more efficient and more profitable. From this perspective, the point is more telling. The economic unsustainability of many farming enterprises appears to have arisen, more or less directly, out of the pursuit of 'efficiency'. An objective which was almost invariably understood in terms of technological development and modernisation.

Problems of the family farm

For almost a century the family farm has been the key structure of the Australian sugar industry. The flexibility and potential for self-exploitation inherent in a family farming system has a number of distinct advantages which have been significant in allowing the Australian sugar industry to remain sustainable. For example, farms which rely entirely on family labour are well placed to withstand the profound year on year differentials in income created by the extreme volatility of the global sugar economy. Incomes may be very low during some years, but with very few expenses necessary for the running of the farm, especially when any capital projects are deferred, farmers can and do simply tighten their belts and survive until a price upturn increases their incomes (see for example, Friedmann, 1985). However, as is the case throughout Australian agriculture (Hindmarsh, 1992; Lawrence *et al.*, 1992), the future of the family cane farm now appears to be increasingly insecure given the cycles of intensification discussed here. This situation is inherently contradictory. Although the family farming structure has proved, in many respects, to be an advantageous basis for the sugar industry, the farms themselves are constantly being undermined by pressures to increase efficiency. In practice, this frequently involves high levels of capitalisation and progressive increases in the minimum size of viable farm units, both of which have tended to make family farms inviable and unsustainable.

Whilst the family farming structure has certainly been significant in allowing the Australian sugar industry to be sustained for almost a century, the situation is perhaps not quite so straightforward as it may at first appear. One government official provided a very two-edged appraisal of the potential for self-exploitation and flexibility provided by the family farming structure:

> What you do find is that when things get tight, economists and bankers like to think of outgoings as variables and fixed costs and there are no such things. It doesn't hold, everything became discretionary. People seem to be able to find more money from various sources including RAS and including social security and they look to their own resources – selling assets, selling blocks of land, we have certainly seen lots of that. Bearing in mind that the sugar industry is concentrated on the coast and of prime real estate value, what we are seeing is urban encroachment becoming a major and growing problem which is going to effect the viability of some of the mills in the area. What they are doing is chopping off their foot to save the leg. . . . You can't keep doing it.
>
> (Personal communication, DPI worker, Mackay)

What this official is suggesting, albeit implicitly, is that the problem is not simply one of surviving year on year fluctuations in income. Rather the implication is that incomes have tended to fall progressively. In this situation, the family farm (and indeed any other production structure) will inevitably become unsustainable irrespective of whatever measures are taken to offset immediate financial problems.

In practice, however, it has not been just the increasingly tight economic situation which has served to undermine the family farm. For example, intergenerational transfer has tended to create a range of problems. On the one hand, many children of cane farmers appear to be somewhat indifferent to the prospect of taking over the family farm and the risks and self-exploitation which this entails. As one Bundaberg cane farmer suggested:

> A lot of the younger blokes, the sons of the farmers can see that its not a certain future, so they'll go and do something else, whatever town people do. And it's something to do with the type of society we've got now. It's a lot of hard work on a cane farm, the younger ones use their brains a bit, go to college, and get themselves a cushy job in an office and get three times the pay. They can see that. Its better than seven days a week, daylight to dark.
>
> (Personal communication)

A Mackay cane farmer evaluated the situation in these terms:

> They have jokes in these Canegrowers things we get, there was a joke in one: the kid was playing up, and the dad said if you don't stop playing up I'll leave you the farm. There is a lot of them like that. People didn't want their kids to stop on the farm because they had had such a hard life. The tendency is for the kids to go away, and the farmers are getting older and older. The younger ones can't afford to buy the farm. The older ones don't want to make them stay.
>
> (Personal communication)

Even where children are committed to the farm, however, there are a range of practical problems. Because most farms are of such a size that they can only provide sufficient income to support one family, established farmers tend to be reluctant to transfer ownership to the succeeding generation until they are able to provide some form of retirement income for themselves. Thus children are often obliged to obtain employment off the farm until well into middle age, by which time they may not wish to return to farming. Similarly, the relatively small nature of many farms means that partible inheritance has not been a realistic option for most cane farmers. In practice, a significant number of farming families have been more or less obliged to adopt a strategy of expansion to allow either single or multiple children to work on the farms whilst the parents were still too young to retire. Such strategies have often proved to be problematic for the families concerned. Many families, especially those who have attempted to expand their holdings rapidly, have experienced severe financial problems. Borrowing to finance the purchase of new land in order to incorporate children into the family business appeared to have been a major cause of unsustainable debt in both Mackay and Bundaberg.

The problems associated with intergenerational transfer of cane farms are reflected in the distorted age structure which exists within this sector. One survey conducted in 1992 identified the mean age of cane farmers in Central Queensland as 56 years, with less than 10 per cent being under 30 years of age (Gray *et al.*, 1993:43). Such a structure is seen as being problematic for a number of reasons. Not the least of these is that the long-term sustainability of the industry would seem to be dependent on the effective reproduction of the ownership structure and labour force. In practice however, figures regarding age structures on cane farms may be misleading. For example, it is common for the father to retain official title to a property long after the effective running of the farm has passed to his children.

The impacts of deregulation

In evidence to the Sugar Industry Commission, growers' representatives argued that deregulation was likely to result in farmers being disadvantaged because of the effective monopoly which mills have in particular cane growing areas. Conversely, the milling sector expressed concern about the power of organised grower groups (Sugar Industry Commission, 1992). Other submissions to the Commission suggested that continued regulation was necessary

> to ensure orderly expansion that is within the capacity of the industry infrastructure and does not threaten the position of those within the industry . . . to provide a means of co-ordinating the scheduling of harvesting and delivery operations in a manner which is equitable between growers . . . to protect growers, potential growers and millers from investment decisions that may not be viable . . . to protect mills from the threat of closure through significant volumes of assignment being transferred out of the mill area; and to preserve and increase industry per unit returns.
>
> (Sugar Industry Commission, 1992)

In practice, almost everybody involved in the Queensland sugar industry appears to be concerned about the possible effects of the current deregulatory programme. At one level, the very idea of deregulation is seen as highly questionable. The Chairman of the Bundaberg Canegrowers described the views of his members in this way:

> We see no advantages in deregulation for us. The sugar industry in Australia certainly was the best organised agricultural industry in the world in terms of how it ran and pulled together. Even though

we fought with the millers we were damned well organised and even when we had a row we had a central board which settled the row. The structure was there to solve the problems as it went along, internally without government sticking their nose in.

(Personal communication)

However, the basis of most considered opposition to deregulation lies in the contention that the nature of sugar production is such that the industry requires regulation for reasons which would not apply in the general case. The suggestion being that the functional interdependence of different sectors of the industry necessitates a high degree of co-ordination between different stages of the production process. And that the inherently unequal relationships which exist between different sectors of the industry is hardly a suitable basis for self-regulation. Certainly, various sectors of the industry believe that their self-interests may be prejudiced by the deregulatory process. It is widely recognised that deregulation will almost inevitably result in some form of restructuring within which there will be both winners and losers. As one DPI worker observed:

What you find when people go bust is that there is someone walking in behind them to take their place. And in so much as that's rationalisation that's probably seen as a good thing by a lot of people . . . in this particular area you will find that those who are in a position to expand are all for deregulation and some of the members of the Canegrowers executive are the people who are in a position to benefit from it the most.

(Personal communication)

An important point here is that various assets and property rights within the industry will almost inevitably be devalued as this restructuring takes place. As we will assess below, the fact that there will almost inevitably be winners and losers in this process provides an interesting commentary on the idea of sustainable development. Whilst this process may allow the industry as a whole to remain sustainable, it will in all probability result in some elements of the current industry becoming unsustainable. Thus it raises key questions about just what should be sustained and what is expendable. As one Canegrowers officer in Mackay put it 'do you want to keep the industry alive or the people within the industry?'.

Coping strategies: the struggle for sustainability

The sustainability of the Australian sugar has clearly embodied a process of contradiction and struggle within which various problems and crises continue to emerge. In this sense the industry has remained sustainable because, thus far, it has managed to address emergent contradictions more or less adequately. The struggles experienced within the farming sector over the last ten or 15 years exemplify this interpretation. Many cane farmers in both the case study locations and indeed throughout Australia have found the sustainability of their own enterprises increasingly prejudiced by a range of developments. For most Australian cane farmers, sustaining anything, simply staying on their farms, has been an ongoing struggle.

One of the most common responses to 'difficult events' amongst Central Queensland farmers during recent years has been 'do nothing' (Gray et al., 1993:48). In some respects, this has been an effective strategy. At least, inaction did not accentuate the problems faced by increasing debt burdens. However, it is clear that in the longer term, farmers who do not respond positively to changing conditions will not remain viable (Lawrence, 1987). Farmers have found themselves on the escalator of the technological treadmill whether they like it or not, and doing nothing has not been an effective long-term strategy. Not least because the minimum viable size of a cane farm is continuously increasing. Of those farmers who did respond positively to the problems they encountered in recent years, three basic strategies predominated: diversification, intensification and the development of off-farm incomes.

One of the major differences in the coping strategies adopted by cane farmers in the two areas studied is that whilst diversification was a common strategy amongst farmers in Bundaberg it was virtually non-existent in Mackay. To some extent it may be that geographical factors played a role in creating this difference: Bundaberg is closer to the major urban markets, but it seems unlikely that this can offer a full explanation. One Mackay cane farmer outlined his own position in these terms:

> They kept saying get a little bit bigger or get out – now they didn't just say that to the sugar industry, they said that to the dairy industry as well, encouraged the little dairy farmer to expand. In the years that it worked they thought it was OK. But then when it didn't work they started to say diversify. They brought in this word diversify. Don't keep all your eggs in one basket. Try and get some more farm income by getting something else going. Some farmers tried that and they went by the wayside as well. They tried aloe vera, they tried all sorts of things. I didn't try that at all, we are 60 kilometres from Mackay and it's hard, when the economy was bad there weren't a lot of jobs or money around anyway.
>
> (Personal communication)

Although most Bundaberg cane farmers seem to have been reluctant to move away from traditional monocultural practices, significant amounts of land were nevertheless taken out of sugar cane during the 1980s. The harvested area fell from almost 40,000 hectares in 1982 to just over 35,000 hectares in 1992 (Canegrowers local records, Bundaberg). A small proportion of this loss, particularly around the city, went into housing developments, but most occurred as farmers diversified into 'small crops' – most commonly, tomatoes, zucchini and peppers. A small number of local cane producers diversified very successfully into small crops in the early 1980s. When incomes from sugar fell later in the decade, a large number of other cane farmers sought to follow this example. However, a glut of production resulted in falling prices for these crops and many new producers found small crops even less profitable than sugar cane. In practice, diversification most usually occurred as a panic response to low returns and pressing cash flow problems. For the most part, it was poorly planned, poorly capitalised and tended to make a bad situation worse for the individual farmers concerned.

Further intensification was the most widespread and common active response to the cost price squeeze which developed during the early 1980s. Several factors underpinned this trend. The difficulties of the 1980s were preceded by a short period of particularly high returns and this may have influenced many farmers, but perhaps more significant was the way in which most farmers seem to have been profoundly constrained in their thinking by the ethos of modernisation which pervades the industry. The overall philosophy of the industry, and successive Australian governments, has always been basically productionist, and the industry has also always prided itself on, and made a considerable virtue of, its technological efficiency. Over and above these factors, the nature of the assignment system has always been likely to promote intensification. On the one hand the regulatory system predicates against intensification in that farm peaks impose a quota on production. But given the situation which has often prevailed in practice where 'over peak', or 'number two pool', sugar has often received a price as high or even higher than that for number one pool sugar, there has been every incentive for farmers to exceed farm peaks. This was a particularly rational strategy given that the majority of production costs are essentially fixed, and where the assignment system made expanding the area under sugar cane difficult. Accordingly, the assignment system often meant that intensification was one of the very few options open to cane farmers wishing to increase their incomes.

The appropriateness of intensification strategies, now at least, appears to be questionable. They are associated with a range of environmental and agronomic problems. And, as events have shown, this sort of strategy has resulted in many farmers accentuating rather than ameliorating their financial problems. It is then curious that faith in modernisation hardly seems to have diminished amongst those farmers who remain in the sugar industry. 'Farm development strategies' and 'property management plans' (terms widely used by various extension agencies in Queensland, see for example, DPI, 1994) have been largely determined by the almost universal perception that

progressive efficiency gains are fundamentally necessary. Farmers often suggest that difficulties arise when enterprises try to expand or modernise too
quickly and thus get into debt, but they hardly question the appropriateness
of the modernisation process itself.

Historically pluriactivity was not common amongst Queensland sugar
cane farmers. The single exception to this being that a significant number
of farmers developed agricultural contracting businesses within the sugar
industry. To some extent this situation may simply reflect the very limited
range of economic opportunities which exist in rural Queensland. Equally,
however, there appears to be a quite considerable social stigma involved in
admitting to needing a secondary income source. By 1992, however, the
situation had changed somewhat and around 75 per cent of Central Queensland cane farms had some source of off-farm income, although of those with
an income only about 50 per cent amounted to more than A$20,000 per
year. The median off-farm income amounted to A$24,000, some A$6,000
more than the mean farm income deficit at this time (Gray et al., 1993:42).

By the late 1980s a significant proportion of Queensland cane farming
families had become heavily dependent on some form of social security provision. The principal support mechanisms which did exist were administered
under the Rural Adjustment Scheme. Jointly funded by the Commonwealth
and State governments, this scheme was designed to 'assist eligible farmers
to improve the productivity, sustainability and profitability of their farming
enterprise' (DPI, 1993). Eligible farmers may receive a range of grants to
facilitate farm improvement programmes. And under some circumstances
such as 'a prolonged severe drought or substantial commodity price falls',
several support measures, including a 'household support scheme', were
available. The scheme also provided 're-establishment' grants for farmers
whose properties were no longer considered viable. All of these support
measures were means-tested and subject to a range of other quite severe
eligibility criteria, and in practice offered only very limited help to most
struggling cane farmers.

A more important factor has been the mutual support provided within
farming communities, particularly within the various ethnic groups which
exist in some areas. Such support takes several forms including assistance
with work on the farm and financial support through informal loans, etc.
According to one industry professional, whilst the attitudes to farming
commonly adopted by some groups within the sugar industry are highly
significant, there are signs that these are changing:

> They had family labour, unpaid family labour. While the traditional
> Anglo-Saxon cane farming family were using employed labour the
> Maltese community often used the extended family. One family in par-
> ticular comes to mind, they were buying farms during the '80s. They
> bought, they did very well. More outgoings become discretionary,

there's more flexibility. But having said that I think that even within
the Maltese community they are becoming more Anglo-Saxonised in
their attitudes and ideals and for at least some of the next generation
those principles might not hold.

(Personal communication, DPI extension worker)

A model sugar industry, a model of sustainability?

The Australian sugar industry is widely held to be a model of efficiency and,
tacitly at least, of sustainability (DPI, 1994). Certainly the Australian industry does not appear to be experiencing problems so profound or so immediate
as those which have affected some sugar producers such as in Barbados.
However, our analysis reveals that the sustainability of the sugar industry is
increasingly tenuous and uncertain. Moreover, it is also apparent that the
sustainability of the sugar industry itself has been dependent on mechanisms
which have tended to produce a range of materially and morally unsustainable outcomes both within the sugar sector and outside it.

Throughout the twentieth century, the Australian sugar industry has
needed to address a series of tensions and contradictions which have periodically threatened dysfunctionality and unsustainability. These contradictions
have been confronted in various ways. They have been postponed through the
more or less constant adoption of new technology; they have been absorbed
through the self-exploitation of small farmers; and they have been held in
check through the development and operation of a comprehensive system of
regulation. But in each of these cases, the resolution of specific problems has
tended in turn to produce new contradictions and new sources of potential
dysfunction. Sustainability for the Australian sugar industry has been and
remains a constant struggle to stay ahead of the game. Certainly this is how
the situation is widely perceived within the industry. Specific problems are
addressed, more or less objectively, as they arise, but whilst the measures
which come into place may be effective in counteracting specific elements of
dysfunction, new contradictions continue to emerge. From this perspective,
the Australian sugar industry is trapped in a process which in the final analysis cannot be sustainable – a road which gets ever steeper and more intract-
able the further one goes along it. And moreover, not only is the process itself
always likely to become untenable, it is by its nature a process which will
always tend to produce a range of materially and morally unsustainable outcomes as the contradictions which continue to emerge become ever more
acute and adequate responses necessarily become ever more exploitative.

Within the industry itself the greatest threat is seen to lie in its high
degree of exposure to the depressed and volatile prices of the global sugar
economy, and the fact that Australia has to compete internationally with a
host of foreign sugar industries where wage costs are often much lower and
which are often heavily subsidised. A particular problem for the Australian

industry is that while most of its competitors still have low wage economies and still benefit from direct subsidies, many are nevertheless becoming increasingly sophisticated in their farming systems. Thailand, which competes directly with Australia in the regional market, is a case in point here. The Thai sugar industry has expanded rapidly in recent decades often using technology developed and produced in Australia (ABARE, 1991).

In practice, the financial viability and thus the sustainability of the Australian industry is doubly threatened by the nature of the global sugar economy. On the one hand, the low prices at which sugar is traded internationally clearly prejudice profitability. But beyond this the volatility of the market, typified by short booms and then long periods of very low prices is also problematic. The structure of the Australian sugar industry, which involves a large number of what are in effect small farms and a milling sector composed of relatively small companies and co-operatives, is well suited to withstanding short periods of low returns, but as recent events have demonstrated, few of the enterprises involved have the resources to withstand long periods of severely depressed prices. In practice, many Australian cane farmers have gone out of business in recent years.

Most of the individual farming enterprises which have failed within the sugar sector have done so because they have acquired unsustainable levels of debt. In itself, debt is not necessarily unsustainable. Where money is borrowed to finance investment which will lead to future productivity gains and enhanced profitability, debt can be seen as positive. Certainly such debt oils the wheels of capitalism throughout the world – credit defers one form of unsustainability. Conversely, however, where debts are accrued because of the operational unprofitability of an enterprise, they are clearly unsustainable in anything other than the very short term. Recent events in the Australian sugar industry blur this distinction. On the one hand, most if not all sugar cane farmers have been adversely affected by the particularly long period of severely depressed sugar prices which began in the early 1980s and extended into the 1990s. Even well-run farms were hard pressed to make any kind of operational profits during this period. That said, there is considerable agreement within the industry that those most severely affected, those who have actually become economically unsustainable, are for the most part those who had borrowed most to invest in new machinery and, allied to this, to increase the size of their holdings. This is in itself somewhat ironic given the philosophy of modernisation which pervades the industry. Certainly, many cane farmers are now very bitter that they were positively and aggressively encouraged to borrow heavily by both the government and the banks during the early 1980s. 'Get big or get out' was very much the industry watchword of the early 1980s, not least because this philosophy was actively promoted by government, quasi-governmental agencies and the financial establishment. Within this, the banks involved in the sugar sector were prepared to extend, and indeed did extend, unrealistically high levels of credit to large numbers of cane farmers. Subsequent events have clearly indicated that the prudence of this agenda was, at best questionable. Many individual cane farmers found what would in any event have been a difficult situation grossly exacerbated by inappropriate development strategies and investment decisions. In practice, these development strategies were effectively defined by the establishment and fundamentally enabled by the financial sector. For a great many small sugar cane farmers the modernisation roller coaster has gone off the rails. However, thus far at least, the industry as a whole has been sustained. Moreover, while many individual farmers may have found the modernisation, techno-fix culture to be unsustainable, various chemical companies, machinery manufacturers and downstream sectors of the industry have clearly benefited from this agenda (Vanclay *et al.*, 1992).

To date, the Australian sugar industry has been able to maintain some comparative advantage by remaining at the forefront of innovation in sugar production. Yields have increased, unit production costs have fallen and the increasing adoption of irrigation has reduced some elements of risk and uncertainty. In the end, however, the modernisation process is inherently unsustainable, at least it cannot be sustained indefinitely. Involved in a competitive global economy the Australian sugar industry has faced and will continue to face constant pressures to improve its efficiency, productivity and cost effectiveness – not so much to improve its profitability *per se* but rather simply to remain economically viable. Thus far the industry has managed, just, to remain internationally competitive. It could be argued that this constant pressure has produced a highly efficient industry. And in some respects this is so – yields are as high as anywhere else in the world and production costs are low. But this type of efficiency hardly equates to sustainable development. Subordination to a technologically driven process of development is always likely to involve a range of unsustainable outcomes. The impacts of the modernisation and intensification of Australian agriculture in general are well documented; see, for example, Watson (1986) and Campbell (1989). The key environmental impacts associated with the sugar industry include: water mining and salinisation problems associated with irrigation practices (Watson, 1986; Hungerford,1987; Williamson, 1990; BSES, 1992); soil erosion (ABARE, 1991); nutrient leaching (ABARE, 1991); eutrophication and, it seems, damage to the Great Barrier Reef (CSIRO, 1990).

A more profound problem lies in the distinct possibility that the modernisation process itself may be fundamentally unsustainable. It requires constant technological innovation, and whilst the mechanisms through which this is achieved are largely institutionalised in Australia, for example through the BSES and the DPI, it is at best an act of faith as to just how long incremental efficiency gains can be made. This must be so because there are absolute limits on just how productive agricultural land can be. Given the relative sophistication of the methods now employed by Australian industry it is highly likely that these limits are already being approached. It is certainly the case that whatever the absolute limits to productivity might be, a process of diminishing returns becomes increasingly significant the nearer these limits are approached. This may appear to be a somewhat trivial point in that it would be possible to expand the area under cultivation as productivity limits were approached on particular farms. And in practice, this seems to be what has occurred as individual sugar farms have tended to increase in size. However, it is not clear whether this is a valid refutation of the argument, either for the sugar industry being considered here, or in the general case. Within the sugar sector at least, geography plays an important role because the extensification of many sugar producing areas is constrained by the lack of suitable land or water supplies, and because the need for cane land to be close to a mill restricts expansion of particular mill catchments. In the more general case, there are other contradictions. For example, production is always going to be limited by levels of consumption if by nothing else.

The fact that the modernisation process predicates progressive increases in farm size may also promote new contradictions and new tendencies to unsustainability. The system of family based sugar cane farms in Australia emerged out of the unsustainability of the plantation system which preceded it. Once established, the family cane farm has, in many respects, proved to be a particularly effectual basis for the industry. Perhaps the most useful quality of the family sugar cane farm has been its potential to absorb contradictory and dysfunctional developments. In large part, this potential stems from the degree of self-exploitation which individual farmers are willing, and in practice have often found themselves obliged, to endure. Over and above this factor, however, a whole range of other informal support mechanisms based around the family farm have also served to maintain the sustainability of individual enterprises and, in a cumulative manner, of the industry as a whole.

The family cane farm is now, however, an endangered species. Since the 1980s, there has been a reduction in the 'independence' of the family farm. They have come to be as much tied to finance capital as to agri-business. And, by the early 1990s, a third phase of highly 'subsumed' farms linked to a more deregulated production system was beginning to emerge. However, the combinations of market and regulatory mechanisms which disenfranchise small farmers not only produce morally unsustainable outcomes, they also constitute a transformation to a new structure of production which in many respects may be inherently less sustainable than that which preceded it. Modernisation has benefited some sectors of the industry largely at the expense of individual cane farmers, but should the family cane farm prove to be unsustainable, the restructuring this would entail may well create new and profound forms of dysfunction.

This dialectic provides a particularly telling commentary on neo-liberal approaches to sustainable development. The demise of the family farm and its replacement by larger production units arguably represents a transition to new structures which by virtue of the fact that they have replaced the old are, by their nature, more efficient. Thus, in one sense, this constitutes 'development': the industry is more productive and more efficient; overall welfare is increased; and the industry remains capable of participating in the global economy. But this interpretation may be flawed. Not least because the transition is not simply the product of the market. In practice, the nature of development has been determined, in part at least, by the nature of the regulatory system and within this by the character of particular institutions such as the BSES. Beyond this, however, the strategies adopted by powerful groups up- and downstream of the agricultural sector also appear to have been extremely significant. Chemical companies, machinery manufacturers, sugar millers and refiners and particularly the banks have all influenced the actuality of sugar cane agriculture. A key point may be that the strategies adopted by these groups appear to have involved more than attempts to promote profit maximisation *per se*.

This situation is inherently contradictory. On the one hand the viability of the family farm has been increasingly prejudiced by 'development' within the Australian sugar sector. However, while the general mode of development existing in Australia since the 1970s has tended to undermine the viability of the family farm, there have also been a whole range of mechanisms put in place to support the individual cane farmer. In practice, the whole regulatory framework has protected the interests of the farming sector. Certainly this was the rationale behind its original development during the early part of the twentieth century. Within this, a variety of governmental and quasi-governmental institutions such as the DPI and the BSES, which for the most part are staffed by plant biologists, geneticists, soil scientists and hydrological engineers, and have performed specific support functions. However, it is clear enough that the remit and the agendas of these institutions have extended beyond any singular concern for the family farm itself. In practice, these institutions became so embedded within a particular interpretation of what development is, and what this means in Queensland, that they no longer really served the interests of the small sugar cane producer. What has been ignored are the structural effects on small farmers and the decline of the small producer as thresholds for extensive and intensive development have increased.

An interesting question arises here as to whether the modernisation ethos which has fundamentally influenced the nature of sugar cane agriculture in Australia is the only practical development option. Certainly it is possible to envisage other possibilities such as, for example, a low input–low output system within which yields would be reduced but so would costs. As such a system would reduce overheads and operational costs it might well have some advantages for farmers operating in a situation where incomes are uncertain. Similarly, there would appear to be some advantages in more poly-cultural production systems. These would be positive in terms of maintaining soil quality and fertility and in preventing the build up of pests and diseases which are a particular problem in sugar cane agriculture. It may be that such alternatives would not be any more viable than those currently favoured. Perhaps the key point, however, is that for most farmers these have never really been considered as options. In practice, the context in which Australian cane farmers have operated has embodied a range of constraints and enablements which have served to promote particular types of development.

The low sugar prices of the 1980s and the cash flow problems which ensued have accentuated and effectively brought to a head what was in any case a progressively unsustainable situation for the family farm in the Australian sugar industry. However, while the problems which many farmers faced were frequently profound and immediate, they could only respond to their situations within the context of the actual and perceived opportunities and constraints which existed. In practice, the responses which farmers actually adopted were clearly influenced by the nature of the institutional and cultural context in which they operated. On the one hand, farmers' perceptions of what opportunities existed were largely conditioned by the institutionally promoted modernisation ethos. But, even within this context, it is clear that the degree of freedom available to individual farmers has been very limited. Many farmers have been fundamentally dependent on, and constrained by, the financial sector. Moreover, the agricultural sector as a whole has become increasingly subordinated to upstream and downstream sectors of the industry. Individual farmers have been influenced by both a culture which centralises a particular type of development and by the progressive impotence of their position.

Regulation as a cause of unsustainability?

The restructuring of the sugar industry into a system based on large numbers of small family farms allied to relatively small local milling enterprises around the turn of the century negated earlier forms of contradiction and dysfunction, but in achieving this it created the potential for new forms of contradiction and crisis. For most of the twentieth century, this potential for dysfunction has been addressed through a purposively constructed regulatory system which controlled almost every aspect of sugar production and therefore arrested the eventual demise of the small producer. However, the regulatory system itself has now come to be seen by government, if not by all those involved in the industry, as a source of dysfunction and unsustainability.

In practice, few of those actually involved in the Australian sugar industry are unhappy with the current regulatory arrangements. Some dissatisfaction has been apparent with respect to specific controls. For example, with those controls which meant that buyers and sellers could not determine the sale price of designated cane land for themselves, and indeed these elements of regulation were apparently often circumvented by under the table payments.

Similarly restrictions on the ways in which co-operative mills can raise investment capital are also seen as being increasingly inappropriate. But these perceived shortcomings relate to specific and relatively minor aspects of the overall regulatory framework. The vast majority of those involved in the Australian sugar industry, and the farmers in particular, believe that the regulatory system as a whole has been for the most part positive and effective. Few understand the rationale for deregulation and many are concerned with the consequences which this process might have for them as individuals within the industry – quite possibly with some considerable justification.

The general argument in favour of liberalisation is not so much that the industry cannot be regulated effectively in the sense that it cannot be managed, rather that this particular type of regulation is now not conducive to 'development' because it tends to inhibit innovation and dynamism. In itself, this is interesting from a sustainability perspective. For example, the basis of this argument implies that an industry such as this cannot be sustained in a steady state, and accordingly that the viability and effectively the sustainability of such an industry depends on its ability to change quite radically. Not least because economic sustainability in the global market place depends upon radical reorganisation of established regulatory structures.

Deregulation of the sugar industry will inevitably produce quite significant restructuring. Under the highly regulated system, mills could not compete with each other and so size was not a major factor. In a deregulated situation, however, whilst individual mills are geographically constrained in the extent to which they can compete for cane suppliers, they are very likely to have to compete for market share when it comes to selling their produce. This may well serve to disadvantage the agricultural sector in several ways. Most obviously mills, now exposed to a competitive environment, will seek to reduce their operational costs – the most significant element of which is the prices they pay their suppliers of sugar cane. Many mills have already begun to embark on this process, for example, by changing to continuous crushing – a development which places an added burden on cane farmers because such a system obliges them to work very extended hours during the harvest season. A competitive sugar economy within Australia is also likely to involve different sugar producing regions and their associated mills competing with each other for market share. Within this, it is clearly possible that some regions will lose out. Particularly so because the current structure of many relatively small milling enterprises is almost inevitably going to be replaced by a structure involving a small number of powerful milling concerns, many of which may well be owned by transnational companies such as Tate and Lyle.

Whilst companies such as Tate and Lyle are averse to becoming directly involved in sugar cane agriculture because of the risks and uncertainties involved, it is also the case that they have no particular interest in sustaining sugar production in any particular region of Australia, or for that matter in

Australia at all. Tate and Lyle are primarily interested in sugar refining and marketing, that is where the greatest value adding and profits are to be found. Where the raw sugar comes from is a very secondary consideration in a situation where at a global scale there is considerable structural oversupply.

Relational unsustainability? Exigency, expediency and expendability

In the Barbados case it was argued that both the unsustainability of the sugar industry and a range of unsustainable practices and events stemmed more or less directly from the unsustainability of extant socio-economic formations. The situation in Australia seems quite different. Certainly, there is no plantocracy, no elite landowning class whose economic base and status are threatened. There are, however, close parallels. The family farming system, although it hardly constitutes an elite class, is nevertheless part of a particular socio-economic formation, and a formation which is also becoming increasingly prejudiced by a range of emergent contradictions.

In the Barbados case, we have seen that an elite group within a particular socio-economic formation became increasingly threatened by the unsustainability of the formation as a whole, and that this particular group was able to maintain its own status, but only through processes which tended to produce a range of unsustainable outcomes. Forms of 'relational unsustainability' were translated into various forms of 'material unsustainability'. The largely insignificant was translated into the consequential. In the Australian case, the formation is different, but the mechanisms involved and the outcomes produced bear close comparison. Here, the formation concerned extends beyond the agricultural sector to include functionally integrated sectors both upstream and downstream of farming. It is the agro-industrial and financial complex in which smaller and smaller numbers of family farmers are located. To some extent, the same could be said about the Barbadian formation, but in this case the integration is more essential and more influential. Indeed, in Australia the farming sector has tended to become an increasingly marginalised, dependent and impotent section of the industry. Economic benefits, power and the potential for self determination have ebbed away from the sugar farming sector, and as in Barbados, when the sustainability of the formation as a whole has become threatened, those with most to lose, those with the most power, have sought to defer the unsustainability of their own positions through whatever means are possible. And again as in Barbados, this has resulted in a range of materially and morally unsustainable outcomes in Australia.

Throughout the last 100 years Australian sugar cane agriculture has progressively become more dependent on higher levels of inputs: chemicals, machinery and capital. The dysfunction and unsustainability which is now emerging within the industry clearly threatens the viability of the enterprises supplying these inputs: enterprises which in any event are always going to face pressures to expand their own production and sales irrespective of developments within other parts of the sugar economy. The development of a high input–high output agricultural system has clearly benefited the upstream sectors of the sugar industry. A key point here is that the conditions which promoted this type of development have become largely institutionalised in Australia. Whether this is by accident or design, the agricultural sector has been conditioned to a particular type of farming – a type of farming which is perhaps not the most appropriate or the most environmentally or socially sustainable, certainly it is only one of a range of potential options.

Modernisation may have allowed the Australian sugar industry to remain internationally competitive, and within this it may have allowed some individual farmers to remain viable. But the most telling commentary on this process comes from an appreciation of the strategies which have been adopted in response to the increasingly acute contradictions which emerged during the 1980s. Here the whole of the industry, including upstream sectors, became increasingly threatened. And whilst farmers' responses varied, the majority have sought to address their difficulties by further intensifying production. This further intensification of an already heavily chemicalised production system clearly has significant implications for material sustainability, particularly in a climate such as that in Queensland where water resources are crucial both within agriculture and outside it. In practice, there is evidence that problems such as nutrient leaching, salinisation and water mining are being accentuated. Perhaps the most widespread unsustainable outcome, however, has been social. Large numbers of individual farmers and their families have suffered prolonged and often quite severe hardship. This has resulted not simply from a period of low incomes due to depressed sugar prices. In practice, many individual farmers, commonly using borrowed money, invested in higher and higher levels of inputs. This clearly served to postpone the unsustainability of those who supply these inputs, but it has proved to be a profoundly unsustainable strategy for many of the farmers concerned. Moreover, social hardship did not result in reductions in the intensity of farming, quite the reverse.

Whilst the chemical companies and machinery manufacturers have clearly benefited from the modernisation process which has taken place within the Australian sugar industry, the process itself has only been possible because of roles played by various government and quasi-governmental institutions and crucially by the banking sector. Institutions such as the BSES and the DPI were quite unambiguous in promoting modernisation, but most individual farmers were only able to embark on such a programme with the support of their banks. In fact, the role of the banking sector is particularly interesting here. On the one hand it involves an input little different to any other used in sugar cane agriculture, but beyond this it has also played a crucial enabling role with respect to much of what else has occurred.

The position taken by the banking sector is clearly questionable in a number of respects. Certainly it resulted in a great many bad debts and a great deal of suffering for many farming families. And with hindsight, it would seem that many of the lending criteria adopted by the banks were manifestly inappropriate. In Queensland many cane farmers found themselves in a situation where they could not service their debts and where their indebtedness far exceeded the value of their properties. Debts have commonly been rescheduled and often continue to be serviced of a fashion, but in many cases with little real chance of them being repaid in the foreseeable future. It certainly seems unlikely that any moderate increase in the sugar price would be sufficient to turn things around. Thus there exists a situation where the banking sector, by virtue of the liens it holds on delinquent loans, is the *de facto* owner of much of the land involved in the sugar industry (a situation which is remarkably similar to that existing in Barbados). Any resolution of these problems would almost inevitably involve quite radical restructuring, entailing both a devalorisation of the assets of cane farmers and the injection of new capital into the industry. In practice, the condition of unsustainable debt which pervades much of the industry can only be ameliorated by a restructuring which would involve a move away from the family farm to a new industry structure, presumably one involving much larger cane production units.

In some ways the demise of the family farm presents an interesting commentary on what sustainable development may involve. In so much as the disappearance of the family farm will, in all probability, involve the formation of new larger production units, development in the Australian sugar industry seems to be going full circle. Development has progressed through one increasingly unsustainable formation, the nineteenth century plantation system, to another based on family farms, only to return to something very much like the original within the space of less than 100 years. Perhaps this new formation will itself be replaced by a 'new' structure of small family farms sometime around the mid twenty-first century. This begins to suggest how the material unsustainability predicated by the capitalist dynamic might be averted in practice. A key point here, however, is that despite its apparent efficiency and inherent advantages, the family farming system and the regime within which it has been constituted, could no longer be sustained. By the early 1990s, the Australian sugar industry had reached a point where 'sustainability' depended on radical restructuring.

Real unsustainability? Structures, mechanisms and outcomes

Development within the Australian sugar sector can be analysed using the conceptual framework outlined in chapters 2 and 3. The Australian sugar industry has evolved through two distinct production regimes and a third is now emerging. As both the first and second regimes become stressed through the periodic emergence of various contradictions, measures were enacted in attempts to defer the unsustainability which these threatened to bring about. Although what occurred in each regime is in some respects best understood as a continuous and singular process, it is still possible to identify particular and perhaps telling moments within their progressions.

In figure 7.6 we relate developments in the second, family farm based, regime which began following the demise of the plantations at the beginning of the century to the model developed in chapter 3. In the applied model, *moment 1* might be the failure to reproduce an adequate labour supply which occurred during the 1940s and 1950s. This barrier to the sustainability of the sugar industry was deferred through the wholesale mechanisation and chemicalisation of the industry. However, whilst this may have negated a specific threat to the sustainability of the industry, it has engendered a range of alternative forms of unsustainability which are indicated at point *a* in the graph. These include: agronomic problems such as the increased likelihood of disease; environmental problems such as salinisation and the over-extraction of water supplies; and economic and social problems arising out of the need to finance modernisation strategies. At this point, the agricultural treadmill has begun to turn and is generating its own momentum.

Moment 2 in figure 7.6 might represent the cash flow crisis which faced

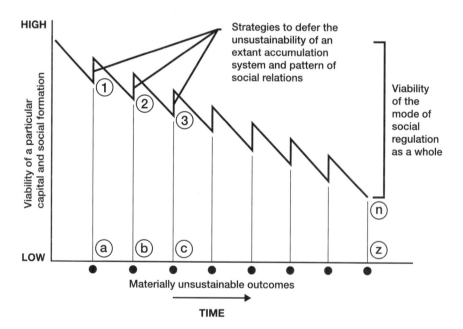

Figure 7.6 The mode of social regulation and unsustainable outcomes

the industry during the 1980s. Responses to this varied, but further modern-isation and intensification was the most common strategy. And again, this produced a range of environmentally and agronomically unsustainable out-comes as shown at moment *b*. Perhaps the most telling outcome here, how-ever, has been the tendency to undermine the family farm as the mechanisms involved produced high levels of debt and increased the size of viable farm units. What this shows is that strategies tend to produce not only materially unsustainable outcomes, but also new and potentially more profound sources of dysfunction and crisis.

Moment 3 shown in the figure 3 might be the deregulatory process which has been enacted from the late 1980s onwards. By this time, the established regulatory system had come to be seen, by the Australian government at least, as being restrictive, inhibitory and a barrier to the efficient modernisa-tion of the industry. Thus it was seen as potentially rendering the industry inefficient and uncompetitive *vis-à-vis* its international competitors and hence unsustainable within the global sugar economy. However, as we dis-cussed above, while deregulation may well promote a restructuring of the industry to a form which is more 'efficient' – at least in terms of global markets, this will clearly involve a range of unsustainable outcomes. These outcomes are likely to be at one and the same time both materially and morally unsustainable, and such that they will promote new barriers to the future sustainability of the industry as a whole. In all probability, deregula-tion will lead to even further intensification of production systems which in turn is likely to exacerbate both agronomic and environmental problems. It is also likely to increase pre-existing pressures on individual cane farms. This is not only morally and socially unsustainable, it is likely to negate the one feature which, more than any other, has allowed the industry to remain sustainable for so long. The theme consistently emerging here is that the intensification of materially unsustainable outcomes is being legitimated by new strategies to defer the unsustainability of the most powerful interests within the established regime. The need, from an analytical perspective, is to understand specifically, and substantively, what mechanisms have been used to do this, and what conditions within the broader mode of social regulation have legitimated and empowered them.

In this sequence of moments within the development of the Australian sugar industry we have a series of actions which were more or less effective in addressing specific elements of contradiction and dysfunction. These actions averted the immediate unsustainability of the industry. However, in each of these cases, the sustainability of the sugar industry and the socio-economic formation which it constituted was only achieved at the cost of a range of materially and morally unsustainable outcomes. And, moreover, in each of these cases new and potentially more profound sources of unsustainability were engendered.

Each of these moments are different and yet they are the same. They are different in that they are specific, but they are hardly singular. All are related to an omnipresent tendency for the industry and the socio-economic formation in which it is constituted to become progressively more unsustainable. Each specific moment of contradiction is just that: a moment in this process. Similarly, the outcomes engendered are contingent and may vary from place to place and at different times, but elements of commonality exist. On the one hand, the outcomes tend themselves to be either materially or morally unsustainable. Beyond this, they are all partial and incomplete solutions to the industry's problems. In solving specific problems they give rise to new and ever more profound problems. Thus development will inevitably arrive at the moment shown as n in figure 7.6. At this moment the cumulative contradictions which beset the industry are too profound to be addressed and the formation as a whole becomes unsustainable. Here, the capitals and patterns of social relations which constitute the formation are threatened with devalorisation. Or more succinctly, this is a point where the sugar industry is no longer capable of supporting their value and validity. A key point here, however, is just how profound the cumulative effects of the materially unsustainable outcomes shown at z are likely to be by this point. Indeed, given the nature of a mode of social regulation which prioritises and ascribes flexibility to economic criteria, it is highly likely that it will be material forms of unsustainability which ultimately prejudice the sustainability of the socio-economic formation. If this is the case we have a mode of development which is intrinsically conditioned to materially significant forms of unsustainability. Moreover, the evidence here also supports the contention that the process will tend to be cumulative. The longer a particular regime is sustained in the face of progressively profound contradiction, the more destructive the outcomes realised will tend to be.

In the Australian case, both the farming sector itself and the upstream and downstream elements of the formation will be extremely vulnerable at this point. There are two general possibilities as this point is approached. First, increasingly exploitative practices can continue until the ecological and social basis of the industry is totally devalued. Second, the patterns of social relations and the capitals employed within the industry can be devalued before this point is reached. The devalorisation of the nineteenth century plantation system is a good example of how devaluing elements of a particular formation can allow a newly structured industry to emerge and remain sustainable. Restructuring the pattern of social relations and property rights which existed at the end of the nineteenth century allowed the industry to be sustained. Similar restructuring might be sufficient to allow the industry to be sustained without recourse to overly exploitative and degrading practices on other occasions.

The restructuring process which is now taking place in the Australian sugar sector will allow the industry as a whole to be sustained through mechanisms which, amongst other things, devalue family farms. But this is not a form of devalorisation which will avert further material unsustainable outcomes. What is being devalued here is the lives of farmers and their families who are little more than a proletariat. The interests of capital are not being prejudiced, rather the reverse. Family farmers are being dis-enfranchised in order that existing finance and industrial capitals and power structures can be sustained. The necessary forms of devalorisation are more radical than this.

As was the case in Barbados, events in Australia can be seen as involving a process through which the unsustainability of some elements of a particular socio-economic formation is averted through mechanisms which translate largely inconsequential 'relational unsustainability' into various forms of more consequential 'material unsustainability'. Whilst the processes and mechanisms involved here do not always act directly to cause unsustainable outcomes, they tend to create conditions in which such outcomes are always likely to be realised. For instance, the promotion of irrigation in semi-arid areas such as exist in some parts of Queensland will tend to promote salinisa-tion. In realist terms, whether or not salinisation actually occurs may be dependent on contingent conditions. And indeed where irrigation has been adopted in Queensland salinisation has not always occurred. But it has oc-curred in some locations. Similarly, the increasingly intensive use of chemical fertilisers has not always created problems with run-off, but it has on some occasions. And again whilst modernisation strategies have not resulted in all the farmers concerned experiencing profound financial difficulties, a great many farmers certainly have. Thus it seems that whilst modernisation may well have served the individual interests of some in the industry it has also involved the promotion of a range of unsustainable events. One of the destructive features of the recent deregulation phase is that it further devalues smaller scale family farming and the regulatory structures on which it is based. It is a shift to an exchequer-led rather than production-led regulatory system.

The overall 'mode of development' which has existed in the Australian sugar industry has tended to produce a range of unsustainable outcomes. But, in practice, specific unsustainable outcomes have normally been addressed as just that: specific, material, discrete problems. Indeed a whole institutional structure, including the DPI and BSES, exists in Australia to do precisely this. This is clearly inadequate from a sustainability perspective. Treating the symptoms is never likely to cure the disease or in the end save the patient. Specific regulatory measures, be they environmental legislation, a social security safety net, or the institutionalised development of new pro-duction technologies will not produce sustainability. Approaches which attempt to address sustainability directly in this way are limited in their scope and may even be counter-productive. Potentially more effective approaches need to understand the causality of unsustainable practices and events more fully and at transcending scales of analysis. In particular, they need to be based on an appreciation of how unsustainable modes of development are able to achieve social and political legitimacy.

Unsustainable practices and events therefore can only usefully be understood as the outcomes of social and economic processes. From this perspective, what actual events come about depends on: the mechanisms involved; dynamic contingent conditions; and on whether or not the mechanisms involved are activated. The Australian sugar case demonstrates quite clearly how a particular mode of social regulation can condition the nature of development in a particular way. In practice, the institutions, values and norms of behaviour which form the mode of social regulation constitute the constraints and enablements which serve to activate the mechanisms which are present. They do this in a biased and selective fashion and because of this they condition the nature of actual events. The mode of social regulation existing in Australia has ascribed priority and flexibility to a particular object of regulation and it has predicated an unsustainable pattern of development involving a range of specific materially and morally unsustainable outcomes. It has conditioned development to the unsustainable. An important point here is that the frequency and profundity of unsustainable outcomes will both tend to increase the longer a particular regime is sustained. Such progression is integral to the rationale and logic of regulation theory. It is also demonstrated clearly enough by events within the Australian sugar industry.

Summary

For almost a century the Australian sugar industry operated as a more or less stable formation. But now it seems that this formation is increasingly precarious. A particularly severe and prolonged period of low prices on the world market for sugar has accentuated pre-existing tensions and contradictions, threatening both the formation as a whole and within this the viability of the various sectors of which it is constituted. As particular contradictions have emerged strategies have been formulated and promoted to offset these. Often these have been successful in a very narrow sense. Immediate threats to the sustainability of the industry and at least some of its constituent elements have, for the most part, been averted or postponed. But almost inevitably new and often more profound sources of disequilibrium have emerged. And as the contradictions threatening the industry have become more acute, appropriate responses to these have necessarily become more exploitative. Thus whilst particular elements of the formation have been, temporarily at least, successful in maintaining the viability of their own positions, a whole range of unsustainable outcomes have been promoted and indeed realised.

What has occurred here appears to be very similar to the patterns of development which have been experienced in Barbados — what might have been seen as relatively inconsequential instances of 'relational unsustainability' have been translated into various forms of 'material unsustainability'. The principle mechanism involved here seems to have been the promotion of 'modernisation' within the industry. However, this mechanism is at best a partial and temporary expedient rather than the basis of a truly sustainable industry. If a more sustainable system is to emerge, more radical solutions than the palliative and essentially conservative measures now being enacted are clearly necessary.

The Australian sugar industry is currently undergoing a period of adjustment in which the traditional family based farming system is being replaced by a new structure, involving much larger agricultural units, within a third regime. This transition is not proving to be a costless exercise, involving as it does the promotion of a range of unsustainable events and practices, including the marginalisation and destruction of the family farming system. If sustainability is perceived as a dynamic condition, then it soon becomes clear that elements of extant formations need to be devalued before new and more viable formations can emerge. What both this case and the Barbadian experience seem to show is that in practice this devalorisation tends to be realised in the form of materially unsustainable events. It tends to be the environment and the weakest sectors of society which are devalued. However, what is not so clear is whether this needs to be the case. It may well be that the devalorisation of other aspects of the formation may be equally sufficient. The structure of the formation can change in ways which do not involve the over-exploitation of resources and the promotion of the materially unsustainable outcomes. Certainly, it is possible to envisage transformations in which the material basis of sustainability is not devalued, that is forms of restructuring which involve a transformation of the relational structures of the formation rather than any redefinition of the system parameters or its material basis. The devaluation of the Australian plantation system which occurred in the late nineteenth century clearly shows how this can work in practice. The key point is that new and viable formations can emerge from the old, but only when some elements of the old are devalued. Current modes of social regulation militate against this. Indeed, they are such that they constitute conditions in which it always tends to be the materially significant that is devalued, despite the fact that the devalorisation of much less significant structures within the formation would be sufficient.

In practice, it does seem to be the case that particular socio-economic formations associated with agricultural development tend to be inherently contradictory and crisis prone. Such formations clearly generate disequilibrating tendencies which act as barriers to their own reproduction. However, as they are currently articulated, modes of social regulation serve to preserve such formations *per se* rather than the material basis of future sustainability. Extant modes of social regulation validate and empower mechanisms which sustain the structure of particular formations and within this particular sets of social relations. In doing this they condition development to the unsustainable. Within any particular regime they encourage and enable the adoption of progressively more exploitative processes and thus they create conditions in which materially and morally unsustainable outcomes become almost inevitable. Developments in the Australian sugar sector support this interpretation. For all the espoused concern over the environment, for all the institutions and values which ostensibly ensure ecological and social sustainability, the regulatory regime as a whole is governed by an object quite separate from these concerns. Both the regulatory regime which has operated within the sugar sector and the wider mode of social regulation in Australia have been inherently conservative, they have been uniquely concerned to reproduce rather than transform the basic structures of the sugar sector. Because of this, they have conditioned development towards the unsustainable.

From this perspective, and in line with our central argument, the need is to understand: (a) how and why regimes become increasingly stressed and crisis prone through time; (b) how and why they are sustained to a point where they necessarily involve high levels of exploitation. This sort of understanding can only be based on a specific and substantive appreciation of which mechanisms have been significant, and precisely which conditions have allowed the unsustainable outcomes associated with these to be realised in practice. In the Australian case, modernisation and related mechanisms such as mechanisation and debt have clearly been significant. Crucially, however, the consequence of these mechanisms has been effected by conditions pertaining within the mode of social regulation. A range of values and modes of behaviour within the farming sector, including a largely unquestioning faith in technology and a determination to stay on the farm until the bitter end, have been significant in this respect. Equally, the neo-liberal ethos of government policy and the nature of the financial system have also been significant. The question is, therefore, would policies which sought to purposively redefine such elements of the mode of social regulation be capable of reconditioning the industry to a more sustainable dynamic?

8

THE CONDITION OF SUSTAINABLE DEVELOPMENT

Here we reconsider our conceptual framework and approach to sustainability. We re-evaluate the model (from chapter 3) which links unsustainable outcomes with underlying causal mechanisms. Particular consideration is given to the ways in which these causal mechanisms are 'activated' by modes of social regulation and the ways in which this 'conditions' the nature of 'development'. We then consider how this conditioning might be modified, and what relevance such an analysis may have as to how sustainable development is articulated in the general case. The chapter concludes with an overall evaluation of the approach to sustainability developed here and suggests how this approach might be further tested, refined and progressed.

Transcending the impasse

If development is perceived as a moral concept involving progressive improvements in the human condition, the poverty, deprivation and inequality which pervade the world of the late twentieth century leave little room for complacency. Putting these moral imperatives aside, there are also good reasons to believe that current modes of development are also prejudicing the future of society in ways which are not only material and absolute, but also quite pressing. However, while examples of the morally and materially unsustainable abound, it is all too apparent that most recent attempts to articulate and operationalise sustainable development have proved to be inadequate. How, therefore, can we move beyond this impasse?

Sustainable development has come to be seen as a utopian and impracticable idea not because the concept itself is unimportant, but rather because current approaches to the idea are inadequate to allow it to be achieved in practice. Although the nature of present day modes of development is such that various forms of unsustainability are the norm, as it has been in our case studies of agricultural development, most current attempts to promote sustainable development still focus on the explicit outcomes produced, rather than the underlying causes. Implicitly, most current approaches concern themselves with where some line can or should be drawn. And how it should be policed. Our case studies show that this will not work and is insufficient to ensure sustainable development. The Australian case in particular, where 'sustainability' is now an explicit goal of a range of governmental and quasi-governmental agencies, is a telling example of the deficiencies of what Redclift (1988:638) terms 'environmental managerialism'. The potential of such approaches is manifestly limited. By working backwards from the bottom line of biologically or morally defined sustainability metrics, they fail to respect the multi-dimensional nature of sustainable development. Neither do they provide for the truly integral solutions which this implies. Attempting to define this line more objectively, or searching for new ways of policing the line, will not produce sustainable development. Our aim has not been to retread these well-worn paths. We have not been concerned to define in some more 'objective' way what is or is not sustainable in Barbados and Australia. Rather our analysis started with the premise that this sort of sophistry has not succeeded and cannot succeed. Defining what is or is not 'sustainable' in a particular place and time is practically impossible; this is not a problem, because this sort of 'objectivity' is neither necessary nor appropriate.

The key question for our thesis was not where the line should be drawn but how to understand and promote sustainability without such lines. We see little point in attempting such metrics. Given the nature of development processes, the lines produced will always tend to be crossed wherever they are located. Approaches which begin from this position have the distinct advantage that they largely bypass the need for any quantitative definition of exactly what is and is not sustainable. This is a strength rather than a sacrifice or weakness. From this perspective, sustainable development is, as Yanarella and Levine (1992:770) suggest, more usefully defined in terms of equilibrium and relationships than in terms of 'metrics' or 'limits'. The one existing approach to sustainable development which claims to address the concept in something akin to this manner is that formulated around environmental economics. But this approach is now demonstrably flawed (Redclift, 1987; O'Riordan, 1991; Dickens,1992; Jacobs, 1994). Its ability to assume equilibrium between capital and nature has lost legitimacy. Regulation theory, on the other hand, is centrally concerned with the disequilibrating and crisis prone nature of capitalist accumulation systems (Aglietta, 1979). Insights from regulation theory are potentially informative, not least because they focus on the need to understand why overly exploitative and degrading practices come about and how they are able to achieve their own social and political legitimacy.

Implicit in the critique of extant approaches to sustainable development outlined in chapter 1, was the contention that most of these have proved to be less than adequate because they have remained trapped within a positivist paradigm. We have attempted to contribute to thinking on sustainable development by developing a revised ontological and epistemological framework defined by critical realism. In so much as critical realist thinking was derived from a rejection of the positivist past, it potentially provides an opportunity to construct a more nuanced and powerful understanding of why and how the unsustainable comes about.

From a realist perspective, objects and structures give rise to tendentially expressed mechanisms. These interact with contingent conditions to produce specific events. The former are just as relevant to understanding sustainability as the latter. As the case studies demonstrate, it is possible to seek to prevent unsustainable events at the level of contingency. For instance, in the Australian sugar industry, the assumption is that environmental problems, such as those associated with water extraction and irrigation, can be adequately addressed through a combination of regulatory controls and various technical fixes. This doesn't work either in the Australian sugar sector or more generally. It assumes that the development logic of productionist and globalised agriculture can largely continue in perpetuity. This is a spurious sustainability.

Alternatively, the realist mode of explanation offers the potential to address different moments and link them to levels of causality. Causality viewed in realist terms, posits the possibility of transforming the objects and structures which produce the mechanisms involved. However, as Dickens (1992) suggests effective intervention at this level is 'unlikely'. However, it may not be necessary – the realist mode of explanation suggests other moments where intervention might be possible. Outcomes depend not just on whether a particular mechanism is present, but equally on whether or not that mechanism is activated. In theory at least, it may well be possible to influence actual events by moderating this process of activation. Thus, for example, while capitalism will always produce mechanisms which will tend to promote unsustainable outcomes, it may be possible to prevent these mechanisms being empowered. The regulation of sustainable development is either impossible or ineffectual at both the levels of structure and contingency. It might, however, be possible to regulate the relationship between these levels, for it is this structuration which conditions the actuality of development.

In capitalist societies whether or not mechanisms are activated is largely a function of the prevailing mode of social regulation. By its nature, a particular mode of social regulation will tend to legitimate, empower and activate a particular type of mechanism. Particular modes of social regulation condition development to particular types of outcome, and they underpin both unsustainable and sustainable development trajectories. Modifying this conditioning in ways which will promote more sustainable modes of development may be possible. Thus the question of which types of mode of social regulation 'activate' 'unsustainable' mechanisms becomes highly significant. Purposive modifications depend on an appreciation of what mechanisms are significant and how their expression is influenced by the conditions in which they occur. This kind of understanding can only be achieved through the analysis of past and present day struggles experienced within the mode of social regulation. The research outlined here was largely concerned to 'unpack' such struggles. To better understand how the ways in which they are resolved tends to condition development.

Conditions of unsustainability

The model developed in chapter 3 attempts to link the disequilibrating tendencies of capitalist accumulation systems with a tendency to engage in progressively exploitative and ultimately unsustainable practices. In this model, unsustainability is not considered in absolute terms; rather it is understood as the likely outcome of the dynamics of capitalist accumulation systems. Regulation theory is fundamentally based on the contention that capitalist economies are not equilibrating. This is the antithesis of neo-liberal economic theory which sees markets as an equilibrating process producing patterns of development based on comparative advantage and maximising overall welfare. These tenets are further reflected onto neo-liberal approaches to sustainable development. The logic is simple enough. In the sugar sector, for example, inefficient, uncompetitive Barbadian producers will be replaced by more productive industries in other locations. The market will ensure that sugar is produced in those locations which have the greatest degree of comparative advantage. More sugar will be produced more cheaply and overall welfare will be improved. In effect, more needs will be met – development will have become more sustainable. The case studies of sugar production and the arguments developed here call into question both the probity and practice of such neo-liberal interpretations.

The history and current position of the global sugar economy demonstrates that intervention and protectionism are an omnipresent and seemingly inevitable feature of such economies. They have existed almost since the inception of a global sugar economy in the seventeenth century. They remain of the most profound significance today. GATT and the Cairns Group aside, this is hardly likely to change. Not only do sugar producers formulate policy inappropriately because they respond to short-lived and unrepresentative price hikes; it is also common for non-economic criteria to be employed in policy formation. The Common Agricultural Policy is a good example of this in the developed world. In the South, many countries have developed sugar industries with the expressed intent of import substitution or simply because this is a relatively easy sector in which to become established. Equally, countries such as Barbados have been reluctant to see established industries collapse for a whole range of reasons, many of which extend beyond the purely economic.

The Barbados case represents a good example of how the logic of comparative advantage can become blurred. Historically, conditions in Barbados and throughout the Caribbean were such that this region was particularly well suited to cane sugar production, and the industry thrived for many years. More recently, technological development has undermined this advantage. Conditions in Barbados have hardly changed but the technology and techniques of sugar production have, and the island can no longer produce sugar cost-effectively; better then that sugar should be produced elsewhere and that Barbados should further develop its tourist industry. But the situation is not this simple. On the one hand the island's historical commitment to sugar and the existence of fixed assets on the island mean that any transition to a more appropriate form of development is always going to be traumatic; and this is made even more problematic by the likely environmental consequences of a collapse of the sugar industry. Moreover, the move from sugar has been orchestrated by a small but influential elite group who have been more or less successful in putting their self-interests before 'development' *per se*. The very fact that the transition has, in some ways at least, been orchestrated undermines the logic of market-led development. Certainly the opportunities open to this group, and equally the constraints in which they have operated, have served to delimit the nature of development there, but the precise structure of Barbadian development has been moulded to class interests rather than the general welfare of the island's population. Within this development paradigm, tourism becomes the only alternative industry which the island has the potential to develop and thus, golf courses aside, much of the agricultural land in Barbados is likely to be under-utilised or simply abandoned. Total unproductivity hardly seems to be commensurate with fulfilling human needs – it cannot be an optimum solution given such an objective. Thus while the Barbadian sugar industry is neither viable by virtue of what it is, nor capable of being maintained, its progressive demise hardly amounts to 'sustainable development'.

Unlike the case in Barbados, conditions in Australia are more appropriate to mechanised and globally organised sugar production. Indeed, technological development has effectively served to benefit the Australian industry. In practice, the Australian industry has striven to maximise this comparative advantage through an almost unbridled commitment to modernisation. Even here, however, where environmental conditions and the social and institutional situation cede the industry what appears to be a high degree of competitive advantage, the industry is not only beset by intense economic pressures, it is also becoming more environmentally and socially 'unsustainable'. Not only has the mode of development which has occurred in the Australian sugar industry involved a range of progressively severe environmental impacts, the dynamics of the industry have also tended to create internal barriers to its future reproduction.

For several decades Australia has been more exposed and attuned to the effects of the global sugar economy than any other country; and events in and around the Australian sugar industry cannot be meaningfully interpreted outside this context. Although most explanations of crisis in both Australia and elsewhere have focused, quite inappropriately, on the effects produced by short-term volatility in the sugar price, the *underlying downward trend* in the sugar price is clear enough and considerably more significant. Sugar prices have, in real terms, been falling more or less consistently at least since the mid-nineteenth century. The dramatic fall in prices engendered by the development of new production technologies and the European beet sugar industry during the last century were initially ameliorated by the transition to mass consumption which Mintz (1985) describes. This option of expanded consumption is no longer available within the global sugar economy – consumption is only likely to increase in the South and then only moderately. In practice, cane producers both in Queensland and throughout the world have to compete for their markets within a context where structural over-production in a competitive economy places constant downward pressure on the sugar price and where the only potential to ensure a cost competitive product lies on the 'supply side'. This has created a situation in which incremental efficiency gains have become a fundamental component of economic sustainability, at least for those production spaces outside the areas of preferential quotas.

This process is 'unsustainable': (a) because it cannot continue indefinitely; and (b) because of the environmental and social impacts it engenders. It is analogous to the rhetoric of 'sustainable economic growth' which has become the watchword of governments throughout the world, and just as growth of any other kind is logically unsustainable, growth in the sugar industry or elsewhere cannot be achieved indefinitely. In practice, static demand has meant that 'economic growth' in the sugar sector has effectively meant seeking ever greater efficiency and this has necessitated the adoption of production methods which are increasingly exploitative and damaging to the wider environment. However, despite the adoption of ever more 'efficient' production systems, the Australian sugar sector has now reached a point where a more radical restructuring of the industry has become almost unavoidable. The social and moral consequences of this restructuring are likely to be profound enough, but an even more telling point lies in the way in which this process is undermining many of those central features of the established system which allowed it to remain sustainable throughout most of this century. Even if the environmental and social contradictions currently emerging within the Australian sugar industry are not profound enough to render the industry totally dysfunctional, the progressive achievement of efficiency gains will always reproduce current patterns of overproduction. As these case studies show, market-led development tends to involve fewer producers producing sugar within narrower margins. When this process is repeated throughout an economy, the ability to consume sugar or any other commodity is always going to be compromised by the fact that not enough people have the means to consume the sugar however cheap it may be. The actuality of this is manifest in the dozens of Southern sugar producing countries where poverty remains the norm (see Coote, 1987). This would have little significance if the sugar sector were unique, but surely it is not. The sugar sector may have singular characteristics, but the logic of the dynamic which defines its momentum is far more general across other food commodities.

The global sugar economy simply does not function in a manner which maximises overall welfare. Rather it serves to undermine the sustainability of sugar production even in the most appropriate locations. Sugar is cheap and getting cheaper (although it is still not cheap enough for many millions in the South), but the mechanisms which have depressed the price of sugar on the world market have led more or less directly to a whole range of increasingly exploitative practices. Although modern production systems may indeed be intrinsically more efficient than those which they replace, the gains achieved are never enough. In practice, technology and modernisation are no more than temporary expedients; they are merely part of a process which requires ever more exploitative forms of accumulation. While the unity of the mode of social regulation which legitimates and enables the mechanisms and practices involved remains intact, increasingly profound outcomes will continue to be realised in practice. Whether or not the practices involved or the specific events they have produced are 'unsustainable' in a particular location and time is hardly significant. What is important, and what is clear enough, is that the progressive nature of this process means that sooner or later these practices and events will transgress into the unsustainable however the notion of 'sustainability' is defined.

The problems of the world's sugar industries are not simply problems of market failure or of protectionism, support for domestic industries or the discounting of indirect or future costs. In practice, the capitalist dynamic effectively necessitates the adoption of increasingly exploitative practices and hence it tends to engender outcomes which are variously environmentally and morally unsustainable (see Marsden, 1997a). In Guyana, Brazil, the Philippines and throughout the South, landless labourers toil in cane fields for a few dollars per day. In Australia, farmers mine water, degrade the environment with chemicals and exploit themselves and their families as they struggle to remain viable in the face of increasingly inauspicious conditions. In Barbados and on other Caribbean islands, potentially productive land is abandoned and left to erode. Capitalist development may or may not be equilibrating. What is certain, however, is that its momentum currently produces a whole range of unsustainable practices and events.

The perspective adopted here therefore requires that the notion of development itself, not just sustainability, is critically reconsidered. Development is usually taken to be a progressive concept implying increased human well-being. This is surely the sense in which it is used in the idea of 'sustainable development'. It is somewhat ironic that such an idea has come to the fore at just the time of new patterns of globalisation of capitalist relations, which economically make it all the more unachieveable (see McMichael, 1996b). There is a problem here of conflating development, which is progressive in this sense, with transformations which simply reproduce key structures within a capitalist mode of production. This ambiguity is reflected in the inconsistent and often discordant ways in which the term sustainable is used both in theory and practice. As the evidence here has demonstrated, even within one sector – sugar – notions of economic, social and environmental sustainability are used collaterally with insufficient consideration being given to just what should be or can be sustained. Within the sustainability literature (Redclift,1991), and in practice, it is often assumed that sustaining each of the economic, social and environmental dimensions of the concept is a prerequisite to achieving sustainable development in some holistic sense. The analysis here questions the logic and propriety of this assumption.

By their very nature of what they are, capitalist modes of development are dynamic and transformational. That is not to say that such dynamism is necessarily progressive. As the model developed in chapter 3 suggests, attempts to sustain economic growth and the validity of particular patterns of social relations are linked, more or less directly, with increasingly pro-found forms of exploitation which have been legitimated within modes of social regulation. This interpretation is supported by the case studies, and is hardly in doubt in the more general case. What remains to be seen is whether or not this is unavoidable.

Whether or not one seeks to rationalise this in Marxist terms, through concepts such as a tendency to overaccumulation or a falling rate of profit, particular capitalist formations, particular capitals and class structures, do tend to move towards dysfunction and crisis. Certainly, this has been clear enough in the case studies included here. Capitalist socio-economic forma-tions are insecure and ephemeral by virtue of the competitive conditions in which they occur and the exploitative basis of what they are, and thus they need to be 'sustained' through various combinations of legitimation and coercion. Dynamism and transformation are engendered not simply through apparently exogenous factors such as technological development, but also through internally generated contradictions. These contradictions may be 'managed' for a while by the prevailing mode of social regulation, but ulti-mately they necessitate change. Historically this process of change has involved a range of materially and morally unsustainable outcomes because of the ways in which emerging contradictions have been addressed. Development has thus been conditioned to the unsustainable.

The constant emergence of contradiction and potential dysfunction in both Barbados and Australia demonstrate that while the contradictions which have emerged have been addressed in ways which often appear at least to be more or less effective, these 'solutions' have been partial and temporary. In practice, the sugar industries in Barbados and Australia have remained operational and the patterns of social relations which they support have been reproduced through mechanisms which serve to externalise contradictions rather than through measures which address them directly. Contradictions have been exported either geographically or temporally rather than negated. This process cannot be sustainable either environmentally or socially.

When the early planters in Barbados encountered problems of severe soil erosion, they invented the cane hole. Soil erosion was checked, but the large amount of labour needed to prepare the land in this way engendered new problems, new sources of dysfunction. These new contradictions were addressed through the use of slave labour. Again this proved to be an incomplete solution as slavery became an increasingly dysfunctional and ultimately an illegitimate and untenable basis for production, and by the late twentieth century cultivation methods have returned to those used in pre-slavery times and soil erosion is once again a problem. In Australia, a variety of pressures have served to prejudice the economic and social basis of the cane farming sector, and whilst farmers have attempted to maintain their viability through mechanisation and chemicalisation, it now seems that the days of the family cane farm are numbered. This is a situation which is both ironic and tragic. The irony lies in the fact that the current restructuring of the Australian sugar industry appears to involve a return to a structure involving fewer but larger farming units. In effect, it is a return to a structure very similar to the plantation based system which existed in the nineteenth century – albeit technologically rather than labour dependent. The tragedy is twofold. On the one hand, a return to a plantation system hardly seems to be progressive and certainly the restructuring is unfortunate for many present day farming families. But perhaps even more tragic than this is the fact that the struggles through which these farming families have sought to protect their positions and the value of their assets have involved both a high degree of self-exploitation and practices which have had progressively severe and by almost any standards unsustainable impacts on the environment.

Analysed in this manner, these cases studies support the general model based largely on theoretical categories derived from regulation theory outlined in chapter 3. Certainly, the emergence of contradiction in the Barbadian and Australian sugar industries appears to reflect Moulaert and Swyngedouw's (1989) regulationist categorisation of crisis. There are 'short conjunctural crises' which can be offset by measures such as technological development or through the acquisition of new markets. These measures are inherently conservative in that they serve to reproduce the extant socio-economic order. There are also more basic crises which require qualitative changes in the accumulation process. The restructuring involved in this sort of change is potentially, at least, much more radical in that established capitals, property rights and patterns of social relations become vulnerable at this stage. In practice, however, as the case studies demonstrate, the types of shift which occur in these circumstances often involve the reproduction of established structures rather than any truly radical transformations.

If we accept that general crisis in a regime of accumulation is necessarily constituted in dysfunction within particular sectors, the recent histories of both the Barbadian and Australian sugar industries would seem to verify the existence of both of these types of crisis. In both of these cases, various forms of contradiction have tended to emerge more or less consistently. Up to a point, it has been possible to address the problems produced through minor adjustments which have served to maintain the formational status quo. Eventually, however, in both of these cases a point has been reached where incremental technological changes have become inadequate to maintain the viability of the industries and a period of restructuring became inevitable. Recent events in Barbados and Australia would also appear to support the contention that the measures which have been employed to avert both types of crisis have tended to involve increasingly exploitative and ultimately unsustainable practices. In this sense, a direct relationship exists between the nature of the global sugar economy and a whole series of materially and morally unsustainable outcomes in rural Queensland. Applying a realist methodology suggests that this relationship can be confirmed and better understood through a process of retroduction which focuses on the causal mechanisms involved.

In the Australian case, two of the key mechanisms involved have been modernisation and debt. In practice, modernisation has been no more than a mechanism through which production could be made more 'efficient'. A falling global sugar price has obligated 'leaner', more 'efficient', production and in practice this has effectively meant the adoption of ever more exploitative production techniques. The outcome of this has been a range of increasingly profound environmental impacts and the progressively severe exploitation and eventual disenfranchisement of a large number of farmers. A clear link also exists between the modernisation process and the pervasive and frequently overpowering levels of debt which exist within the Australian sugar industry. It is easy enough to understand why individual farmers faced with unserviceable levels of debt tend to resort to pragmatic measures with scant regard for their long-term consequences in often desperate attempts to meet their immediate commitments. However, most of this debt reflects not simply the particularly low sugar prices of the 1980s, but also the recent purchases of technology and land made by farmers who accepted the need to become more efficient in order to remain economically viable. Somewhat ironically, the high and untenable levels of debt which abound amongst Australian cane farmers are a key factor underpinning the current phase of restructuring. Certainly, unserviceable debt burdens make farmers more likely to lose their land to larger and better capitalised concerns, and accordingly growth in farm debt fuels the concentration process. Restructuring can make the industry viable again, at least in the short term, because a new equilibrium is achieved as some aspects of the established industry structure are devalued. In practice, however, it is the livelihoods of the farmers and their

214

families which are being devalued rather than any of the more powerful elements of the established formation.

In so far as the modernisation process has been forced upon the Australian sugar industry by a progressively falling sugar price, it is also clear enough that the dynamic and transformational nature of the global sugar economy has been a significant causal factor underpinning unsustainable events in Queensland. Thus it might well be argued that a 'substantial' or 'internal' relationship exists between the nature of the global sugar economy and various unsustainable events. However, the situation is not so straightforward. Actual events in Queensland also reflect the fact that the mechanisms involved here were socially and politically activated. It is not simply the nature of the global sugar economy or mechanisms such as modernisation and debt which are unsustainable. Those elements of the mode of social regulation which 'selected' these mechanisms are equally significant and problematic.

In Barbados, many unsustainable events, such as accelerated soil erosion, have been underpinned by the recent large scale flight of capital from the sugar sector. In large part, this reflects the penetration of specifically capitalistic accumulation processes, such as tourism and financial services, into the island's economy. On the one hand, the plantation system had clearly become anachronistic and contradictory. For example, difficulties in ensuring an adequate labour force became increasingly problematic as internally generated tensions were complemented by emergent employment opportunities in the tourist sector. But beyond this, it also seems that the incorporation of Barbados within the international tourist market and an increasingly globalised food system have provided new and attractive investment opportunities for the capital previously employed in the sugar sector. In this sense, it was not simply the inherently 'inefficient' nature of the plantation system or the environmental constraints faced by the Barbadian sugar industry which have rendered it dysfunctional. These may well have been significant and conducive factors, but given that the plantation system did produce sugar for over 300 years and the preferential nature of Barbados' EU quota arrangements, they hardly provide a full or convincing explanation. Equally if not more significant than these has been the emergence of new and more profitable sources of accumulation. The problem in Barbados has not been so much that sugar could not be produced profitably (at least within the protected and subsidised context in which it takes place), but rather that it has become relatively unprofitable compared with tourism or overseas investment opportunities. So, in so much as the removal of capital from the sugar sector has been significant in producing a range of environmental and morally unsustainable events, a clear enough relationship exists between these events and the penetration into Barbados of more purely capitalist accumulation processes than had previously pertained.

However, the actual mode of development and the specific 'unsustainable' events which have occurred in Barbados cannot be fully explained in these terms. Actual patterns also reflect the existing power structures on the island and the nature of the mode of social regulation. Just as the nature of the institutions and social values existing in Australia fundamentally validated and enabled the modernisation process and thus had a key influence on the pattern of development there, conditions in Barbados have served to licence and direct the nature of development in that location. The political and institutional conditions and the broader mode of regulation existing in Barbados conditioned development towards particular outcomes. Several factors have been important here. On the one hand, 'regulation' in the narrow sense of the term is not particularly effective on the island. Equally however, the Barbadian government has adopted increasingly liberal economic policies, largely in acquiescence to pressure from the IMF. The establishment of institutions such as the Barbados Securities Exchange, and the now very liberal approach to financial regulation on the island have significant implications. In practice, these factors have allowed an elite group, formally associated with and empowered through the ownership and control of sugar production, to sustain its own position through mechanisms which devalued more materially and morally significant aspects of development on the island. Whilst it is clear enough that the situation in Barbados was such that this type of outcome was always likely, it is equally clear that it was not inevitable.

What emerges in both case studies is not only the 'bounded rationality' of those who face problems and the subjectivity of their strategies, but also the ways in which their responses are selectively empowered or invalidated by the conditions in which they are articulated. On the one hand strategies are conditioned by the fact that emergent contradictions tend to become increasingly profound the longer the established regime is maintained. Thus appropriate responses tend to become more and more exploitative as time passes. Beyond this, the nature of the strategies is also conditioned by the perceived and actual opportunities and constraints experienced by the key actors and groups involved. In Barbados, the economic and political power of the elite group has produced particular outcomes. In Australia the pattern of development has been fundamentally conditioned by the institutionally and culturally defined modernisation ethos of the industry and the broader ideology of deregulation. In both the case study locations, those involved in the sugar industry, planters, farmers, millers, politicians and so on, have all attempted to address specific problems as they have emerged, but the types of strategies which have been devised and which have actually been 'successful' have been largely determined by the institutional and social context – the mode of social regulation – in which they were formulated and promoted. Our analysis of the Barbadian and Australian sugar industries supports the contention that effective policies for the promotion of sustainable development need to be formulated around a more substantive appreciation of how particular modes of social regulation condition the nature of development. Development is currently conditioned to the unsustainable not simply because it occurs in an exploitative and disequilibrating global and local capitalist system, but also because the established modes of social regulation ascribe particular and arguably inappropriate and unnecessary priority and flexibility to specific elements of that system. From this perspective, the analytical framework provided by the food systems and food regimes discourses outlined in chapter 5 achieves considerable significance.

In general terms, the contradictions which arise and prejudice established accumulation systems and social structures can be addressed in various ways. Production costs can be reduced and markets can be expanded either geographically or through the provision of credit. Each of these mechanisms is apparent in recent patterns of development and each tends to be closely associated with various forms of unsustainable development. For example, overaccumulated nationally based capitals have been widely translated into international capital – perhaps the final and most destructive of all relational formations. Overaccumulated national capitals, unsustainable in a domestic context, have become international in attempts to maintain their validity through the acquisition of new production and consumption bases. If one considers the history of colonial expansion by the European powers this is clearly neither a unique nor new process. The direct consequences of such neo-colonial expansion have included high levels of Third World debt and in many cases the severe over-exploitation of Southern agricultural resources. In practice, of course, geographical restructuring is in itself only a temporary measure as new contradictions inevitably emerge. A key feature of international capital however is that it is not possible to move again to a larger scale and the only recourse is to more exploitative forms of accumulation within particular places and regimes.

Within these processes neither the environments nor the populations of particular localities hold any particular significance. The demise of traditional industries is in itself stressful, but this pales into relative insignificance as new contradictions emerge within subsequent formations. In these circumstances, practices which are increasingly exploitative of both nature and people appear to be both necessary and appropriate, and thus we witness massive over-exploitation. This is empowered, if not totally legitimated, through mechanisms such as hegemony, international debt and in the extreme case by force of arms. Consider the role of the World Bank and the IMF, the GATT process or various military interventions in the South. Within this manifest and increasingly pervasive expression of the unsustainable, TNCs and the capitals they embody have for the most part remained viable. But our thesis and analysis suggest that their viability tends to be ensured in ways which undermine the true basis of sustainability. This is not sustainable development. The achievement of sustainable development requires that the reasons for this prioritisation and the conditions which ascribe flexibility to particular and nonessential objects of regulation are re-examined and redefined.

Conditions of sustainability

The picture which emerges from this analysis is one which supports the general model constructed in chapter 3. The analysis confirms the relationship between the capitalist dynamic and a tendency to adopt increasingly exploitative, and ultimately unsustainable practices. The two regional and national socio-economic formations analysed have tended to become increasingly stressed through time, and whilst this has remained possible, inherently conservative measures, including labour coercion, agricultural modernisation, the exploitation of small farmers and protectionism, have been promoted in attempts to maintain the validity of these formations. However, whilst measures of this sort have often been sufficient to reproduce the validity of established patterns of social relations in the short term, they have also tended to degrade either the environment or the lives of some groups in society. Indeed, as the case studies considered here demonstrate, they currently often do both. Moreover, ultimately, measures of this sort have become insufficient to sustain the status quo and a more radical phase of restructuring has become unavoidable. Even here, however, *it has tended to be materially and morally significant environmental and human resources which are devalued rather than the extant relational structures of society*. In this sense, both of the case studies do support the theoretical constructs employed in the original model by suggesting that increasingly exploitative practices will be legitimated and unsustainable outcomes will tend to be realised for as long as the particular mode of social regulation as a whole remains intact. In both of the case studies, the particular institutional and social contexts which served to legitimate and empower the mechanisms involved were important in allowing unsustainable outcomes to be realised. These conditions therefore need to become a central focus of attention in progressing the social science of sustainability.

Key questions for proponents of sustainable development are whether and how it might be possible to effectively promote more sustainable modes of development by purposively modifying specific elements of the mode of social regulation. Viewing sustainability from a realist perspective is essential in that it presents an opportunity to formulate policy in terms of these causal factors. In particular, such a perspective shows the potential importance of preventing unsustainable outcomes through targeting policy on the institutional and social context in which significant causal mechanisms operate. But building on the opportunities which this different approach presents in theory depends on a methodology which can signify quite specifically just which values and institutions need to change and how. This has to go beyond the general suggestion that sustainable development will be built around value shifts in society. In itself, this sort of prescription is neither original nor particularly profound. However, when links have been suggested between capitalist modes of production and unsustainable patterns of development, the analyses which have emerged have tended to be highly generalised and difficult to relate to policy (Murdoch, 1992). The methodology utilised in this project has, at least, begun to show how this sort of discernment might be achieved.

The Australian case study paints a picture in which a progressively falling sugar price has more or less consistently created disequilibria within the sugar sector. In simple terms, farm incomes have failed to stay in line with the costs of production. Up to a point, it has been possible to offset this trend and the unsustainability which it embodies, through a series of incremental efficiency gains achieved largely through the adoption of new technology. This process has allowed the established agro-industrial structure to be reproduced through almost a century. However, a situation has now been reached where further gains of this sort are no longer sufficient and a period of more radical restructuring has been forced upon the industry. In Barbados the situation has been different at the surface level, but the process is essentially the same. There, the conditions of the ACP agreement maintained the effective level of returns to producers, but rising production costs still created a disequilibrated situation. The Barbados case shows that attempts to contrive and manage some form of equilibrium do not work well in practice. The Lomé provisions have hardly created a sustainable equilibrium within the sugar sector and attempts by the Barbadian government to construct such a state through direct support for the industry have also been unsuccessful. Similarly, the profound failure of successive International Sugar Agreements testifies that equilibrium cannot simply be constructed and subsequently managed at a global scale. These failures do not simply reflect the inherent difficulties of managing complex systems of production and consumption; the real problem lies in the fact that such measures attempt to maintain the status quo in a context where this simply is not possible.

This is not an argument in favour of neo-liberal market-led approaches to sustainable development. Indeed it is quite the reverse. The suggestion is not that a perfect market without intervention, protectionism or the support for domestic industries which occurs in the sugar sector would produce more sustainable patterns of development. Rather the point is that the inherently conservative measures which are used to sustain extant patterns of social relations within capitalist economies are key causes of a whole range of unsustainable outcomes. Institutions such as the World Bank, GATT and the Cairns Group may argue that the liberalisation process is commensurate with sustainable development but in practice this is not the case. As the deregulation of the Australian sugar sector demonstrates so unequivocally, such liberalisation implicitly licenses the type of conservative measures which translate the unsustainability of extant class structures into materially and morally unsustainable outcomes. Indeed, in so much as developments in Australia are typical of the more general case, the tendency for property rights to become increasingly concentrated under such conditions makes it all the more likely that unsustainable structures of production will be further extended than would otherwise be the case. The Barbadian experience certainly demonstrates how the congruence of economic and political power can cause this to occur. The point here is, of course, that the longer established regimes are enabled to remain viable the more exploitative the practices they embody will tend to become. The liberalisation discourse becomes the latest means by which such unsustainable social relations and modes of regulation can continue the vulnerable hegemony.

A key issue here is that whilst it is possible to maintain some form of equilibrium in the short term, the measures through which this is achieved cannot be costless. Up to a point, it may appear that equilibrium can be restored through apparently costless exercises, which for example involve expanded consumption, but as the sugar industry demonstrates these are inevitably incomplete and temporary solutions. The relatively short period in which the post-war Keynesian experiment was able to achieve any success suggests that this is also true in the more general case. In the end, mechanisms which are capable of restoring equilibrium within an established socioeconomic structure necessarily involve practices which are more exploitative than those they replace. This is the case whether we are considering modernisation of the Australian sugar industry, the national economies of post-war Europe or the present day global economy. In each of these cases established class and centre–periphery relations have been maintained through mechanisms which are increasingly exploitative. The key point, however, is that equilibrium could be restored through measures which devalued the capitals and patterns of social relations existing in established economies rather than the environmental and human resources upon which they have been based. The case studies provide a useful example of how this can be achieved. Maintaining adequate labour supplies became a problem on the early Australian plantations. In this case, the dysfunctionality which this produced was forestalled by a transition to a production system based on family farms. This transition from plantation-based production to family farms is potentially very significant from a sustainability perspective. The established plantation system had become increasingly stressed and untenable, but the industry was maintained in a relatively benign manner. The environment was not subjected to increasing levels of exploitation; and from a social perspective, many more individuals were able to make a decent living than had previously been the case. The only thing devalued in this restructuring was the capital and property rights of the plantation owners – the exact opposite of what is currently happening in both Barbados and Australia. This example indicates one means through which key environmental and social components of sustainability might be maintained through the transformational dynamic within which development necessarily occurs. In practice, however, this sort of transformation is exceptional. As the case studies demonstrate, even when industries and economies undergo radical phases of restructuring it tends to be the material and moral basis of sustainability which is devalued rather than any less consequential relational structures.

The logic of this is that in a competitive economy established structures are, sooner or later, going to be devalued; and trying to extend their validity beyond a certain point may be politically expedient and attractive to those who have most to lose, but it is profoundly inappropriate from a perspective which values the ecological and social components of sustainability. Sustainable development can only occur, therefore can only be understood, within the transformational context of the capitalist dynamic. The key is to reconsider what should be sustained and what is expendable within this process. From this perspective, measures to promote sustainable development may well need to encourage the devalorisation of established relational structures rather than preserve them as is the case at present. In the end, it would seem, particular socio-economic formations and within these particular capitals and class structures have to be and will be devalued. The questions are when and how, not if. As they are currently constituted, modes of social regulation facilitate, encourage and to some extent determine the processes through which largely inconsequential elements of 'relational unsustainability' are protected. Thus they also tend to predicate the 'materially unsustainable' because the protection of one, almost inevitably, tends to involve the promotion of the other. Hence, sustainable development needs to be constructed within modes of social regulation which do not just incorporate environmental and social criteria. They need to delimit and constrain the flexibility currently ascribed to capital and extant power structures. This 'object' of regulation requires redefinition and expansion.

A key problem here is the fact that modes of social regulation come about and achieve validity through a process of social conflict and struggle rather than through any form of objective promotion. New modes of social regulation cannot simply be constructed as entire and valid wholes. However, the analysis here has begun to show that this may not be necessary. New modes of social regulation cannot be constructed *per se,* but it may be possible to disestablish the priority currently ascribed to sustaining the value of capital and the validity of established patterns of social relations.

In Barbados, struggles experienced within the mode of social regulation were resolved in ways which served particular class interests. In Australia, similar struggles were resolved in ways which benefited particular sectors of the sugar industry. In both of these cases, established relational structures were sustained but only through mechanisms which produced a range of materially and morally unsustainable outcomes. Modes of social regulation with different biases would not necessarily have produced these same outcomes. If the mode of social regulation pertaining in Barbados had not been such that it ceded undue priority to mechanisms which merely reproduced the position of the island's economic elite, patterns of development there might well have been somewhat different to what they are today. Similarly, had farmers in Australia not embarked so vehemently on a process of modernisation and had not incurred the high levels of debt which so many now have, their situation would hardly be any worse than it is now; and many of the environmental impacts which have occurred in recent years might well have been avoided.

In the Australian case, the current deregulatory programme is a good enough example of how bias within the mode of social regulation is socially constructed. It is clear enough that deregulation is an inherently conservative measure which will allow the most powerful elements of the current industry to be sustained at the expense of the most vulnerable. A key point here is that whilst some form of restructuring may have become inevitable, the decision to deregulate the sugar industry was ideologically defined. Notwithstanding an increasingly neo-liberal bias in its own political agenda, the Australian government's vociferous condemnation of intervention and protectionism elsewhere left it with little option but to 'put its own house in order'. The Commonwealth government may have had only limited room for manoeuvre, but alternatives did exist. In practice, deregulation was determined as much by ideology, political pressures and expediency as by any determinism inherent in the nature of the context in which the sugar industry operates.

Modes of social regulation are not socially constructed *per se,* but they are clearly imbued with bias not just in Australia but throughout the world. Currently, this bias is generally associated with a perceived need to reproduce the value of capital and existing power structures and environmental and social goals are marginalised. However, as this research has demonstrated, the achievement of sustainable development depends just as much on constraining the flexibility of capital as it does on prioritising environmental controls or legislation to protect the most vulnerable in society. There is a fundamental need to re-evaluate current systems of property rights and particularly those elements of regulation which facilitate the mobility of capital.

In Barbados the sugar industry has clearly been prejudiced by a whole range of contradictions, but the unsustainability of the current mode of development has been profoundly influenced by the ways in which capital has become freer to move both within and out of the island. In Australia, the industry has clearly been prejudiced by a falling sugar price, but the plethora of environmentally and socially unsustainable outcomes which occurred have been specifically conditioned by the wider mode of social regulation and the increasingly neo-liberal position of the Australian government. This is also true in the more general case. The political agenda throughout the world is increasingly one which seeks to sustain the status quo through the adoption of progressively exploitative practices at a global scale. This is not sustainable. The institutions and values which legitimate this process are the basis of unsustainability. Changing these is a *sine qua non* of a more sustainable future. The radical nature of this agenda may well be the real sustainability impasse.

Beyond the impasse?

The aim of this volume has been to progress the theory and practice of sustainable development. The need to transcend the limitations and inadequacies of established approaches is hardly in doubt. In particular, the necessity of moving beyond understanding unsustainable events as discrete and unembedded occurrences and attempting to address them as such is clear enough. This volume has attempted to move beyond the current impasse by reconsidering sustainability issues within a realist ontological and epistemological framework using theoretical constructs derived from regulation theory. The results have not invalidated the theoretical potential of this approach. Rather, they have substantiated the model linking the dynamics of capitalist accumulation and unsustainable outcomes developed in chapter 3 of this volume. However, the fact that sustainability debates might be usefully informed by a closer engagement with social theory was never in doubt. The real tests are not whether such engagement can inform policy, but whether social theory needs to be or can be modified in the light of sustainability concerns. Indeed, the problem for this project was not so much linking the dynamics of capitalist accumulation systems with a range of increasingly exploitative practices, rather, the principal difficulty lay in testing and refining this model.

This was always likely to be problematic as the established methodology for 'putting realism into practice' is hardly straightforward. Realist methodology is difficult to articulate in practice, not least because actual events often reflect complex patterns of causality involving a range of contingent factors, plural mechanisms and factors which influence the ways in which mechanisms are or are not activated. Moreover, in practice, it is clear enough that objects and structures other than those considered in this analysis may well have significant causal powers and a more convincing and powerful model might well have included these more fully than was the case here. That said, what was sought here was 'practical adequacy' rather than any totally complete truth. Thus, two key questions arise. First, is the model developed within this volume convincing in that it remains theoretically and empirically sound? And second, is it practically adequate in the sense that it can provide the basis of a productive and useful approach to sustainable development?

Established realist methodology suggests that an initial 'model' developed around theoretical constructs and actual events is refined through a reflexive process of testing, substantiation and modification. The original model used

223

here was constructed largely around theoretical categories defined by regulation theory. The case studies explored in this project appear to support the general regulationist interpretation of capitalism as an inherently crisis prone and transformational process within which accumulation systems are necessarily sustained through modes of social regulation. It should be recognised, however, that many of the theoretical constructs employed here are some way removed from the central tenets of established regulationist thinking. As chapter 5 indicates, they have not been applied to the study of food systems or regimes. In particular, an attempt has been made to extend key regulationist concepts which are usually applied at a macro scale to individual sectors. The problem here is that regulationist analysis, in the food case as elsewhere, is usually concerned with a unity which necessarily extends beyond any individual sector. Although it might well be argued therefore that it is inappropriate to apply the precepts of regulation theory to the small scale and the unique, this does not undermine the validity of the approach adopted here. Certainly, the established socio-economic formations in both case study locations appear to tend to be unsustainable because increasingly profound contradictions keep emerging. In themselves, these might not be the same macro scale sources of dysfunction on which most regulationist analysis has focused, but they are surely reflections of these.

The analysis also supported the conceptual model in that relationships were established between the capitalist dynamic and what by any reasonable definition would constitute unsustainable practices and events. In both the case studies significant relationships and causal mechanisms were identified. The inherent 'unsustainability' of established capital and class structures in both Barbados and Australia was postponed, but only through recourse to ever more profound forms of exploitation. In practice, environmental and human resources have been degraded while established capitals and patterns of social relations have been reproduced. And in both the case studies, the prevailing modes of social regulation have ascribed priority and flexibility to mechanisms which preserved the value of capital and protected established patterns of social relations whilst marginalising and devaluing environmental and moral components of development.

The notions of 'formational' and 'material' sustainability employed throughout this analysis retain some analytical utility. Certainly these conceptual categories can be applied relatively convincingly in the case studies. More significant than these, however, are the related contentions that sustainability can most usefully be conceived of in terms of equilibrium and that it is most properly understood as a condition rather than some quantifiable demarcation of the nature–society relationship. For the most part, the analysis of the case studies also confirms the relevance and probity of these conceptual categories. The case studies support the contention that current modes of social regulation tend to condition development in ways which make unsustainable outcomes the norm. Thus the analysis here also suggests that potential for positive change lies in understanding how this conditioning might be objectively changed. The approach and methodology employed here can take thinking on sustainability beyond vague and unembedded notions of institutional and value change in society, allowing a more objective determination of just which institutions and values need to change and what form these changes need to take.

This research has just begun to clarify how the specifically realist notion of mechanisms being 'activated' can form the basis of a new approach to sustainable development. In Barbados, the strategies pursued by the 'plantocracy' would not have been effective if they had not been legitimated and empowered by a particular set of institutions and social values. In Australia, neither the whole emphasis on modernisation nor the current deregulatory programme could have been enacted without a mode of social regulation which legitimated these actions. In both of these cases it seems to be clear enough that the prevailing modes of social regulation have conditioned the nature of development.

The deregulatory programme in the Australian sugar sector and the liberalisation of financial regulation in Barbados are relevant not simply in their own right, but also because they mirror wider trends in a situation where increasingly neo-liberal agendas affecting food as well as other sectors have gained prominence throughout the world. This raises the important question of whether the general conclusions reached from the analysis of the case studies embody any wider significance. There is no evident reason to suppose that there is anything singular about the sugar sector which should restrict the relevance of any conclusions reached here. Certainly this sector has particular characteristics as do both of the case study locations, but both the sugar sector and the locations studied are far from unique. The fortunes of both are tied directly to the global sugar economy and whilst actual outcomes may vary from place to place, many of the processes and mechanisms which affect these locations are the same as those which affect sugar producers throughout the world. Equally, whilst much has been written concerning those characteristics of agriculture which differentiate it from other forms of capitalist production, these do not preclude a more general applicability for the analysis here, particularly as sugar production is, in any case, an agro-industrial process (see Drummond 1996). Beyond these points, there also appears to be abundant evidence to support the wider applicability of the general conclusions reached in this volume. Particular socio-economic formations in other sectors and locations appear to be crisis prone and temporary – certainly there is little evidence that any have been sustained indefinitely. Equally, the suggestion that the emergence of contradiction is normally addressed through essentially conservative strategies and that these tend to involve materially and morally unsustainable outcomes can be easily and convincingly transposed to other situations.

More generally, understanding unsustainable development simply in terms of the theoretical categories related to purely capitalist structures clearly paints an incomplete picture. However, it may well be that such categories are pervasive and profound enough to have a very broad relevance. Understanding the relationship between capitalist dynamics and unsustainable outcomes may provide a practically adequate explanation of why development tends to be conditioned to the unsustainable. Understanding how and to what extent this relationship might be modified by purposive social action may well be the key to overcoming the sustainability impasse.

This volume has begun to explore how social theory can provide a conceptual framework and methodology which is relevant and useful to the achievement of sustainable development. It has begun to consider how development is often effectively conditioned to the unsustainable and how this conditioning might be reversed. Clearly, much work still remains to be done. The analysis here has identified what some of the key causal mechanisms in the two case studies are – the mobility of capital in Barbados, and modernisation and debt in Australia. It has also identified a key significance in the particular conditions which activate and empower the mechanisms allowing 'unsustainable' outcomes to be realised. In both of these cases an increasingly neo-liberal bias within the modes of social regulation appears to have been crucial in conditioning development towards the unsustainable. This provides a general commentary on neo-liberal approaches to sustainable development, but if this critique is to be translated into positive action, the conclusions reached in this volume need to be further tested and refined.

The model developed here needs to be applied in other locations and in other sectors in order that a clearer notion of just what conditions are important and just how they affect development can be constructed. Equally importantly, further consideration needs to be given to what modifications are appropriate and possible. The analysis here suggests that modes of social regulation need to be reconstructed in ways which reverse the priority and flexibility currently accorded to mechanisms which preserve the value of capital and the preservation of extant patterns of social relations. Indeed the apparent conclusion is that modes of social regulation should encourage the devalorisation of capital and fixed assets. Whilst it clearly is the case that the potential for effective agency is limited in the sense that new modes of social regulation cannot be constructed as valid wholes, purposive modifications might still be possible. If we accept that increasingly neo-liberal political agendas in both Barbados and Australia have served to condition the actuality of development in particular ways, we must surely also accept that different political agendas might produce different and more desirable outcomes. This conclusion needs more comparable research to establish precisely how this might be achieved. For example, although inflation has for some time been a central concern of some regulationist schools (Jessop, 1990), the links between this mechanism and sustainability issues have not been adequately explored. The tentative conclusions of this research would suggest that in so much as this mechanism does serve to devalue existing capitals and class structures, economic policies which are fundamentally concerned to control inflation may well be quite damaging from a sustainability perspective. Consideration also needs to be given to just how radical and politically inexpedient such an agenda would be, and to whether it would, in practice, be too radical to have any realistic chance of being put in place given the current pervasiveness of neo-liberal ideologies and related modes of social regulation. Related to this, questions regarding the territoriality of regulatory modes, especially at the state level, needs to be further explored in relation to sustainability concerns. And this, in turn, posits key questions about the extent to which modes of social regulation depend on social practices which are not constituted through the state. A modified realist mode of explanation and the sort of methodology applied in this volume can allow progress to be made in answering these questions.

Further engagement with social theory and critical refinement of it in relation to sustainability is the one agenda which will allow both the theory and practice of sustainable development to be progressed. The way ahead will clearly be difficult and it would be quite unrealistic to believe that simple and incontestable strategies for a more sustainable future will emerge quickly or easily. The analysis here has been embryonic and is certainly incomplete, but it has begun to chart a path beyond the sustainability impasse, and it has elucidated a methodology which might be used to further explore this new path.

REFERENCES

Abbott, G. (1990) *Sugar.* Routledge, London.

Adams, W. (1993) 'Sustainable development and the greening of development theory'. In: Schuurman, F. (ed.) *Beyond the Impasse: New Directions in Development Theory.* Zed Books, London.

Adamson, A. (1972) *Sugar Without Slaves: the Political Economy of British Guiana 1838–1904.* Yale University Press, New Haven.

Aglietta, M. (1979) *Theory of Capitalist Regulation: the US Experience.* New Left Books, London.

Agro-Industrial Management Services (AIMS) (1991) *A Management Proposal for the Manufacturing and Marketing Sector of the Barbados Sugar Industry.* Unpublished report.

Allen, J. (1983) 'Property relations and landlordism: a realist approach'. *Society and Space,* 1:2, 191–204.

Alston, M. (ed.) (1991) *Family Farming: Australia and New Zealand.* Centre for Rural Social Research, Charles Sturt University, Wagga Wagga.

Antoine, P. (1989) *Producer Behaviour and Implications for Sugar Policy in the Barbados Sugar Industry.* Unpublished MSc thesis, University of Florida.

Arce, A. and Marsden, T.K. (1993) 'The social construction of international food: a new research agenda'. *Economic Geography* 69:3, 23–311.

Australian Bureau of Agricultural and Resource Economics (ABARE) (1985) *Sugar Industry Working Party Report, August 1985.* Australian Government Publishing Service, Canberra.

Australian Bureau of Agricultural and Resource Economics (ABARE) (1991) *Submission 91.5 to the Industry Commission: The Australian Sugar Industry in the 1990s.* Australian Government Publishing Service, Canberra.

Bager, T (1997) 'Review of the globalisation of agriculture and food'. *Sociologia Ruralis,* 37:1, 80–3.

Bahro, R. (1984) *Socialism and Survival.* Heretic Books, London.

Barbados Statistical Service (1992) *Annual Statistics: 1990.* GOB, Bridgetown.

Barbados Water Resources Group (1978) *Barbados Water Resources Study. Vol. 3. Water Resources and Geo-hydrology.* Ministry of Finance and Planning, Barbados.

Barbier, E. (1989) *Economics, Natural Resource Scarcity and Development.* Earthscan, London.

Barnes, T. (1996) *Logics of Dislocation: Models, Metaphors, and Meanings of Economic Space.* Guilford Press, New York.

Barrett, W. (1965) 'Caribbean sugar production standards in the seventeenth and eighteenth centuries.' In: Parker, J. (ed.) *Merchants and Scholars.* University of Minnesota Press, Minneapolis.

Beckles, H. (1990) *A History of Barbados from Amerindian Settlement to Nation-state.* Cambridge UP, New York.

Benton, T. (1994) 'Biology and social theory in the environmental debate'. In: Redclift, M. and Benton, T. (eds) *Social Theory and the Global Environment.* Routledge, London.

Bhaskar, R. (1975) *A Realist Theory of Science.* Leeds Books, Leeds.

Bhaskar, R. (1979) *The Possibility of Naturalism.* Harvester, Brighton.

Bhaskar, R. (1994) *Plato, etc.* Verso, London.

Blaikie, P. (1985) *The Political Ecology of Soil Erosion.* Methuen, London.

Blaikie, P. and Brookfield, H. (1987) *Land Degradation and Society.* Methuen, London.

Blume, H. (1985) *Geography of Sugar Cane.* Albert Bartens, Berlin.

Bonanno, A., Busch, L., Friedland, W., Gouveía, L., and Minzione, E. (eds) (1994) *From Columbus to Conagra: the Globalisation of Food and Agriculture.* University of Kansas Press, Lawrence, KS.

Booker Tate (1993) *Barbados Sugar Industry Restructuring Plan; Executive Summary and Volumes 1–4.* Unpublished report.

Borrell, B. and Duncan, R. (1989) 'A survey of the costs of world sugar policies'. *World Bank Research Observer,* 7:2, 171–94.

Boyer, R. (1990) *The Regulation School: a Critical Introduction.* Columbia UP, New York.

Bryant, R. L. (1992) 'Political ecology : an emerging research agenda in Third world studies'. *Political Geography*, 11:1,12–36.

Bundaberg Cane Productivity Committee (1993) *Factors Affecting Cane Farm Productivity and Profitability in the Bundaberg District. Preliminary Report.* Unpublished report.

Bundaberg Canegrowers (1991) *Annual Report.* Bundaberg Canegrowers, Bundaberg.

Burch, D., Rickson, R. and Annels, R. (1992) 'The growth of agribusiness: environmental and social implications of contract farming'. In: Lawrence, G., Vanclay, F. and Furze, B. (eds) *Agriculture, Environment and Society.* MacMillan, Melbourne.

Burch, D., Rickson, R. and Lawrence, G. (1996) (eds) *Globalisation and Agri-food Restructuring.* Avebury, Sydney.

Bureau of Sugar Experiment Stations (BSES) (1992) *BSES Annual Report to the Minister for Primary Industries.* BSES, Indooroopilly, Queensland.

Burrows, R. (1989) 'Some notes towards a realistic realism'. *International Journal of Sociology and Social Policy,* 9, 46–63.

Buttel. F. and Gertler, M. (1982) 'Agricultural structure, agricultural policy and environmental quality: some observations on the context of agricultural research in North America'. *Agriculture and Environment,* 7.

Cameron, J. and Elix, J. (1991) *Recovering Ground: a Case Study Approach to Ecologically Sustainable Rural Land Management.* Australian Conservation Foundation, Melbourne.

Campbell, A. (1989) 'Landcare in Australia: an overview'. *Journal of Soil and Water Conservation,* 2:4, 18–20.

Canegrowers (1991) *Bundaberg Sugar Industry.* Canegrowers, Bundaberg.

Caribbean Conservation Association (1991) *Environmental Profile: Antigua and Barbuda.* St. Thomas, Caribbean Conservation Association, Bridgetown.

Caribbean Conservation Association, Island Resources Foundation, United States Agency for International Development (1994) *Environmental Agenda for the 1990s: a Synthesis of Eastern Caribbean Country Environmental Profile Series.* Caribbean Conservation Association, Island Resources Foundation, USAID, Bridgetown.

Carnegie, A. (1996) 'Governmental institutional organisations and legislative requirements for sustainable development'. In: Griffith, M. and Persauld, B. (eds) *Economic Policy and the Environment: the Caribbean Experience.* Centre for Environment and Development CED, University of the West Indies.

Clarke, S. (1988) 'Overaccumulation, class struggle and the regulation approach'. *Capital and Class,* 36, 59–92.

Cloke, P. and Le Heron, R. (1994) 'Agricultural deregulation: the case of New Zealand.' In: Lowe, P., Marsden, T. and Whatmore, S. (eds) *Regulating Agriculture.* Fulton, London.

Cloke, P., Philo, C. and Saddler, D. (1991) *Approaching Human Geography: An Introduction to Contemporary Theoretical Debates.* Paul Chapman Publishing, London.

Cock, P. (1992) 'Co-operative land management for ecological and social sustainability'. In: Lawrence, G. (ed.) *Agriculture, Environment and Society: Contemporary Issues for Australia.* MacMillan, Melbourne.

Collier, A. (1994) *Critical Realism: An Introduction to Roy Bhaskar's Philosophy.* Verso, London.

Coote, B. (1987) *The Hunger Crop: Poverty and the Sugar Industry.* Oxfam Public Affairs Unit, Oxford.

CSIRO (1990) *Australia's Environment and its Natural Resources: an Outlook.* CSIRO, Canberra.

Dale, A., Arber, S. and Proctor, M. (1988) *Doing Secondary Analysis.* Hyman, London.

de Boer, H. (1981) *Assessment of Ratooning Performance after Mechanical Harvesting.* BSIL, Barbados.

de Boer, H. (1994) *Recent Sugar Cane Yields in Barbados.* BSIL, Barbados.

Deerr, N. (1949) *The History of Sugar, Volumes I and II.* Chapman Hall, London.

Department of Primary Industries (DPI) (1993) *The Rural Adjustment Scheme.* DPI, Brisbane.

Department of Primary Industries (DPI) (1994) *Property Management Planning: Planning for Sustainability and Profit.* DPI, Brisbane.

Dickens, P. (1992) *Society and Nature: Towards a Green Social Theory,* Harvester, London.

Dovers, S. and Handmer, J. (1992) 'Uncertainty, sustainability and change'. *Global Environmental Change,* 2:4, 262–76.

Doyal, L. and Gough, I. (1991) *A Theory of Human Need.* Macmillan, London.

Drummond, I. (1996) 'Conditions of unsustainability in the Australian sugar industry'. *Geoforum,* 27:3, 345–54.

Drummond, I. and Marsden, T. K. (1995a) 'Sustainable development: a role for social theory'. *Global Environmental Change,* 5:1, 51–63.

Drummond, I. and Marsden, T. (1995b) 'The Barbados sugar industry: a case study of unsustainability'. *Geography,* 80:4, 342–54.

Drummond, I. and Symes, D. (1996) 'Rethinking sustainable fisheries: the realist paradigm'. *Sociologia Ruralis,* 36:2, 152–62.

English Nature (1992) *Strategic Planning and Sustainable Development*. English Nature, Peterborough.

Fielding, N. and Fielding, I. (1986) *Linking Data*. Sage, London.

Fine, B., Heissman, M. and Wright, W. (1996) *Consumption in the Age of Affluence: the World of Food*. Routledge, London.

Food and Agriculture Organisation of the United Nations (FAO) (1987) *Sugar: Major Trade and Stabilisation Issues in the Eighties*. FAO, Rome.

Friedland, W. (1994) 'The new globalisation: the case of fresh produce'. In Bonanno, A. *et al.* (eds) *From Columbus to Conagra*. University of Kansas Press, Lawrence.

Friedmann, H. (1985) 'Family farming enterprises in agriculture: structural limits and political possibilities'. In: Cox, G., Lowe, P. and Winter, M. (eds) *Agriculture: People and Policies*. Allen & Unwin, London.

Friedmann, H. (1993) 'The political economy of food'. *New Left Review*, 197, 29–57.

Friedmann, H. and McMichael, P. (1989) 'Agriculture and the State system'. *Sociologia Ruralis*, 29, 93–117

Gardner, J. (1990) 'Decision making for sustainable development: selected approaches to environmental assessment and management'. *Environmental Impact Assessment Review*, 9, 337–66.

Girvan, N. (1973) 'The development of dependency economies in Caribbean and Latin America: review and comparison'. *Social and Economic Studies*, 22:1, 1–33.

Goodman, D. and Redclift, M. (1991) *Refashioning Nature*. Routledge, London.

Goodman, D. and Watts, M. (1994) 'Reconfiguring the rural or fording the divide? Capitalist restructuring and the global agro-food system'. *Journal of Peasant Studies*, 22:1, 1–49.

Government of Antigua (1991) *Antigua Environmental Profile*. Caribbean Conservation Association, Bridgetown.

Government of Barbados (1956) *Development Plan*. GOB, Bridgetown.

Government of Barbados (1965) *Development Plan*. GOB, Bridgetown.

Government of Barbados (1979) *Development Plan*. GOB, Bridgetown.

Government of Barbados (1988) *Development Plan*. GOB, Bridgetown.

Government of Barbados (1993) *Development Plan*. GOB, Bridgetown.

Graves, A. (1993) *Cane and Labour. The Political Economy of the Queensland Sugar Industry, 1862–1906*. Edinburgh University Press, Edinburgh.

Gray, I., Lawrence, G. and Dunn, T. (1993) *Coping with Change: Australian Farmers in the 1990s*. Centre for Rural Social Research, Charles Sturt University, Wagga Wagga.

Hannah, A. (1989) *Economic and Social Implications of Sugar Cane Processing in Developing Countries*. International Labour Office, Geneva.

Harvey, D, (1977) 'Population, resources and the ideology of science'. In: Peet, R. (ed.) *Radical Geography*. Maaroufa, Chicago.

Harvey, D. (1989) *The Urban Experience*. Johns Hopkins University Press, Baltimore.

Harvey, D. (1996) *Justice, Nature and the Geography of Difference*. Blackwell, Oxford.

Hayek, F. (1988) *The Fatal Conceit*. Routledge, London.

Healy, P. and Shaw, T. (1994) 'Changing meanings of environment, in the British planning system'. *Transactions of the Institute of British Geographers*, NS 19, 425–38.

Heismann, M. (1993) *The Persistence of Sugar in British Food Supply from 1900 to the Present Day*. SOAS, University of London, Working Paper No. 35.

Hindmarsh, R. (1992) 'Agricultural biotechnologies: ecosocial concerns for sustainable rural agriculture'. In: Lawrence, G., Vanclay, F. and Furze, B. (eds) *Agriculture, Environment and Society.* MacMillan, Melbourne.

Hodge, I. and Dunn, J. (1992) *Rural Change and Sustainability: a Research Review.* ESRC.

Hudson, C. (1987) *The Diversification Story.* Paper presented to 5th Annual BSTA Conference, Bridgetown.

Hudson, C. (1990) *Production of Sugar and Yields of Cane in Barbados: Trends in Recent Times and Possible Explanations.* Paper presented to 8th Annual BSTA Conference, Bridgetown.

Hungerford, L. (1987) *Sugar Cane Farming in the Bundaberg District 1945–1985.* MA thesis, University of Central Queensland.

International Sugar Organisation (ISO) (1994) *Sugar Yearbook.* ISO, London.

IUCN, UNEP, WWF (1991) *Caring for the Earth: A Strategy for Sustainable Living.* IUCN, UNEP, WWF, Geneva.

Jacobs, M. (1994) 'The limits to neoclassicism: towards an institutional environmental economics'. In: Redclift, M. and Benton, T. (eds) *Social Theory and the Global Environment.* Routledge, London.

Jesson, B. (1991) 'Not a government made to last'. *Australian Society*, 2, 26–36.

Jessop, B. (1990) 'Regulation theories in retrospect and prospect'. *Economy and Society*, 19:2, 153–216.

Jessop, B. (1995) 'The regulation approach, governance and post-Fordism: alternative perspectives on economic and political change'. *Economy and Society*, 24:3, 307–33.

Kautsky, K. (1988) *The Agrarian Question.* Zwan, London.

Keat, R. and Urry, J. (1982) *Social Theory as Science.* Routledge, London.

Kerr, J. (1988) A *Century of Sugar.* Mackay Sugar Co-operative, Mackay.

Lance Jones and Company (1975) *The Australian Sugar Industry.* Lance Jones and Company, Brisbane.

Laszlo. E. (1972) *The Systems View of the World: the Natural Philosophy of the New Development in the Sciences.* Brazillier, New York.

Latour, B. (1988) 'The politics of explanation: an alternative'. In: Woolgar, S. (ed.) *Knowledge and Reflexivity: New Frontiers on the Sociology of Knowledge.* Sage, Thousand Oaks, Ca.

Lawrence, G. (1987) *Capitalism and the Countryside.* Pluto, Sydney.

Lawrence G. and Vanclay, F. (1994) 'Agricultural change in the semi-periphery: the Murray–Darling Basin in Australia', in McMichael, P. (1994) *The Global Restructuring of Agro-food Systems.* Cornell University Press, Ithaca, New York.

Lawrence G. and Vanclay, F. and Furze, B (1992) *Agriculture, Environment and Society: Contemporary Issues for Australia.* MacMillan, Melbourne.

Leborgne, D. and Lipietz, A. (1988) 'New technologies, new modes of regulation: some spatial implications'. *Environment and Planning D: Society and Space*, 6:3, 263–80.

Le Heron, R. (1993) *Globalised Agriculture*, Pergamon, Oxford.

Lélé, S. (1991) 'Sustainable development: a critical review'. *World Development*, 19:6, 607–21.

Licht, F. O. (1993) *World Sugar and Sweetener Yearbook 1993.* F. O. Licht, Ratzenburg.

Lovering, J. (1990) 'Neither fundamentalism nor "new realism": a critical realist perspective on current divisions in social theory'. *Capital and Class*, 4:2, 30–54.

Lowe, P., Marsden, T. and Whatmore, S. (eds) (1990) *Technological Change and the Rural Environment. Critical Perspectives in Rural Change: Volume 2*. David Fulton, London.

Lowe, P., Marsden T., Whatmore, S. (eds) (1994) *Regulating Agriculture. Critical Perspectives in Rural Change: Volume 6*. David Fulton, London.

McGregor, A., Baxter, S., Hagelberg, G. and McGregor, M. (1979) *The Barbados Sugar Industry: Problems and Perspectives*. GOB, Bridgetown.

Mackay Canegrowers (1994) *Local Records*. Canegrowers, Mackay.

Mackay Sugar (1994) *Introduction to Mackay Sugar*. Mackay Sugar, Mackay.

McMichael, P. (1994) *The Global Restructuring of Agro-food Systems*. Cornell University Press, Ithaca, New York.

McMichael, P. (1996a) *Development and Social Change: a Global Perspective*. Pine Forge Press, California.

McMichael, P (1996b) 'Globalisation: myths and realities'. *Rural Sociology*, 61:1, 25–55.

Manning, K. (1983) *In Their Own Hands*. Farleigh Co-operative, Farleigh, Queensland.

Marsden. T. (1997a) 'Reshaping environments: agriculture and water interactions and the creation of vulnerablity'. *Transactions of the Institute of British Geographers*, 22:3, 321–37.

Marsden, T. (1997b) 'Creating space for food: the distinctiveness of recent agrarian development'. In: Goodman, D. and Watts, M. (eds) *Globalising Food: Agrarian Questions and Global Restructuring*. Routledge, London.

Marsden, T. and Arce, A. (1995) 'Constructing quality: emerging food networks in the rural transition'. *Environment and Planning A*, 27:8, 1261–79.

Marsden, T., Munton, R., Ward, N. and Whatmore, S. (1996) 'Agricultural geography and the political economy approach: a review'. *Economic Geography*, 72:4, 361–75.

Marsden, T., Murdoch, J. and Abram, S. (1998) 'Rural sustainability in Britain: the social bases of sustainability'. Chapter 31 in Redclift, M. and Woodgate, G. (eds) *The International Handbook of Environmental Sociology*. Edward Elgar, London.

Marsden, T., Murdoch, J., Lowe, P., Munton, R. and Flynn, A. (1993) *Constructing the Countryside*. University College London Press, London.

Meadowcroft, J. (1997) 'Planning, democracy and the challenge of sustainable development'. *International Political Science Review*, 18:2, 167–90.

Meadows, D. H., Meadows, D. L., Randers, J. and Behrens, W. (1972) *The Limits to Growth*. Universe Books, New York.

Miller, G. (1987) *The Political Economy of International Agricultural Policy Reform*. AGPS, Canberra.

Mintz, S. (1985) *Sweetness and Power: the Place of Sugar in Modern History*. Penguin, Harmondsworth.

Moore, D. S. (1993) 'Contesting terrain in Zimbabwe eastern highlands – political ecology, ethnography, and peasant resource struggles'. *Economic Geography*, 69:4, 380–401

Moulaert, F. and Swyngedouw, E. (1989) 'A regulation approach to the geography of flexible production systems'. *Environment and Planning D: Society and Space*, 7:3, 327–45.

Murdoch J. (1992) *Rural Sustainability*. ESRC Countryside Change Initiative Working Paper 32.

Murdoch, J. and Marsden, T. (1995) *Reconstituting Rurality: Class, Power and Community in the Development Process*. University College London Press, London.

Nijkamp, P. and Soeteman, F. (1988) 'Land-use, economy and ecology: needs and prospects for co-evolutionary development'. *Futures*, 20:6, 621–34.

Norgaard, R. (1988) 'Co-evolutionary agricultural development'. *Economic Development and Cultural Change*, 32:3, 606–20.

Nurse, J. (1978) *The Future of Sugar Cane Production in Barbados*. Unpublished thesis, Cornell University.

O'Connor. J. (1988) 'Uneven and combined development: a view from the south'. *Race and Class*, 30:3, 5–15.

O'Riordan, T. (1991) 'The new environmentalism and sustainable development'. *The Science of the Total Environment*, 108, 5–15.

Outhwaite, W. (1987) *New Philosophies of Social Science: Realism, Hermeneutics and Critical Theory*. Macmillan, London.

Owens, S. (1994) 'Land, limits and sustainability: a conceptual framework and some dilemmas for the planning system'. *Transactions of the Institute of British Geographer*, NS 19, 439–56.

Pastor, R. and Fletcher, R. (1991) 'The Caribbean in the 21st century'. *Foreign Affairs*, 70:3, 99–116.

Pattullo, P. (1996) *Last Resorts: the Cost of Tourism in the Caribbean*. Cassell, London.

Pearce, D. and Turner R. K. (1990) *Economics of Natural Resources and the Environment*. Harvester, London.

Pearce, D. (1995) *Blueprint 3: Measuring Sustainable Development*. Routledge, London.

Peck, J. and Tickell, A. (1992) *Local Modes of Social Regulation? Regulation Theory, Thatcherism and Uneven Development*. SPA Working Paper 14, University of Manchester.

Peet, R. and Watts, M. (1993) 'Development theory and environment in the age of market triumphalism'. *Economic Geography*, 69:3, 227–54.

Pierce, J. (1992) 'Progress and the biosphere: the dialectics of sustainable development'. *Canadian Geographer*, 36:4, 306–20.

Powell, R. and McGovern, M. (1987) *The Economic Importance of the Sugar Industry on the Queensland State Economy*. University of New England, Armdale.

Pratt, A. (1995) 'Putting critical realism to work: the practical implications for geographical research'. *Progress in Human Geography*, 19:1, 61–74.

Queensland Sugar Corporation (1991) *Queensland Sugar in Focus*. Queensland Sugar Corporation, Brisbane.

Queensland Sugar Corporation (1992a) *Sugar Notes 1991*. Queensland Sugar Corporation, Brisbane.

Queensland Sugar Corporation (1992b) *Pricing Australian Raw Sugar. Submission to the 1991 Industry Commission Inquiry into Raw Material Pricing for Domestic Use*. Queensland Sugar Corporation, Brisbane.

Queensland Sugar Corporation (1992c) *Single Desk Marketing of Queensland Raw Sugar. Submission to the National Competition Policy Review*. Queensland Sugar Corporation, Brisbane.

Redclift, M. (1987) *Sustainable Development: Exploring the Contradictions*. Methuen, London.

Redclift, M. (1988) 'Sustainable development and the market: a framework for analysis'. *Futures*, 20:6, 635–50.

Redclift, M. (1990) 'The role of agricultural technology in sustainable development'. In: Lowe, P., Marsden, T. K. and Whatmore, S. (eds) *Technological Change and the Rural Environment*. John Wiley, London.

Redclift, M. (1991) 'The multiple dimensions of sustainable development'. *Geography*, 76:1, 36–42.

Redclift, M. (1992) 'Sustainable development: needs, values rights'. Paper presented to the Rural Economy and Society Study Group, Hull University, November, 1992.

Redclift, M. and Benton, T. (1984) *Social Theory and the Global Environment*. Routledge, London.

Rees, J. (1990) *Natural Resources: Allocation, Economics and Policy*. Routledge, London.

Rees, J. (1992) 'Markets – the panacea for environmental regulation?' *Geoforum*, 23:3, 383–94.

Roberts, R. (1992) 'Nature, uneven development and the agricultural landscape'. In: Bowler, I., Bryant, C. and Nellis, M. (eds) *Contemporary Rural Systems in Transition. Volume 1: Agriculture and the Environment*. Wallingford CAB International, Wallingford.

Sánchez, R. (1964) *Sugar and Society in the Caribbean*. Yale UP, New Haven.

Saunders, K. (1982) *Workers in Bondage*. University of Queensland Press, Brisbane.

Sayer, A. (1984) *Method in Social Science: A Realist Approach*. Hutchinson, London.

Secretary of State for the Environment (1994) *Sustainable Development: the UK Strategy*. HMSO, London.

Senate Committee on Industry, Science and Technology (SCIST) (1989) *Assistance for the Sugar Industry*. Australian Government Publishing Service, Canberra.

Shlomowitz, R. (1982) 'Melanesian labour and the development of the Queensland sugar industry 1863–1906'. *Research in Economic History*, 7, 327–61.

Smith, N. (1984) *Uneven Development*. Blackwell, Oxford.

Soil Conservation Unit (1987) *Soil Erosion in the Scotland District, Barbados*. GOB, Bridgetown.

Sparks Companies Inc. (SCI) (1992) *A review of the Barbados Agricultural Sector: Implications for Future Investment*. Inter-American Development Bank, Bridgetown.

Sturgiss, R., Tobler, P. and Connell, P. (1988) *Japanese Sugar Policy and its Effects on the World Market*. ABARE, Occasional paper 104, Australian Government, Canberra.

Sugar Board (1991) *Annual Report, 1990–91*. Australian Government Publishing Service, Canberra.

Sugar Industry Commission (1992) *The Australian Sugar Industry, Report No. 19*. Australian Government Publishing Service, Canberra.

Tate and Lyle (1994) *Annual Report 1993*. Tate and Lyle, London.

Tomich, D. (1990) *Slavery in the Circuit of Sugar*. Johns Hopkins, Baltimore.

Trotman, A. R. (1994) 'Agroclimatic study of Barbados, 1970–1990: rainfall'. *Caribbean Meteorological Institute, Technical Note 25*. Caribbean Meteorological Institute, Bridgetown.

Usborne, D. (1994) 'How big sugar sours the everglades'. *Independent on Sunday*, 30 January.

Vanclay, F., Lawrence, G. and Gray, I. (1992) *Environmental and Social Consequences of*

Capitalist Agriculture in the Australian Context. Paper presented at the 8th World Congress for Rural Sociology, 11–16 August.

Watson, C. (1986) 'Irrigation'. In: Russell, J. and Isbell, R. (eds) *Australian Soils: the Human Impact.* University of Queensland Press, St Lucia.

Watts, D. (1987) *The West Indies: Patterns of Development, Culture and Environmental Change Since 1492.* Cambridge University Press, Cambridge.

Watts, D. (1997) 'Environmental degradation, the water resource and sustainable development in the Eastern Caribbean'. *Caribbean Geographer.* 6:1, 2–14.

Whatmore, S. (1995) 'From farming to agribusiness'. In: Johnston, R., Taylor, P. and Watts, M. (eds) *Geographies of Global Change.* Oxford: Blackwell.

Wheelwright, T. (1990) 'Should Australia be striving to become more competitive in the international market place?' In: Golan, A. (ed.) *Questions for the Nineties.* Left Book Club, Sydney.

Whelan, R. (1989) *Mounting Greenery.* Institute for Economic Affairs, London.

Wickham, O., Smith, L. and Bellamy, S. (1990) 'The effect of longer ratooning on sugar cane yields in Barbados'. *Proceedings of 8th Annual Conference. Barbados Society of Technologists in Agriculture,* 14–17.

Williamson, D. (1990) 'Salinity – an old environmental problem'. In: Australian Bureau of Statistics *Yearbook Australia.* Australian Government Publishing Service, Canberra.

World Bank (1980) *Alcohol Production from Biomass in Developing Countries.* World Bank, Washington.

World Bank (1986) *World Development Report 1986.* OUP, New York.

World Bank (1992) *World Development Report.* OUP, London.

World Commission on Environment and Development (WCED) (1987) *Our Common Future.* OUP, Geneva.

Worrell, L. (1982) *The Economy of Barbados 1946–1898.* Central Bank of Barbados, Bridgetown.

Yanarella, E. and Levine, R. (1992) 'Does sustainable development lead to sustainability?' *Futures,* 24:2, 759–74.

Yearley, S. (1996) *Globalisation and the Environment.* Sage, London.

INDEX